Everyone's Trash

Everyone's Trash

One Man Against 1.6 Billion Pounds

Duncan Watson

Peter E. Randall Publisher
Portsmouth, New Hampshire
2024

©2024 Duncan Watson
All rights reserved.

ISBN: 978-1-942155-77-5
Library of Congress Control Number: 2024911956

Peter E. Randall Publisher
5 Greenleaf Woods Drive, Suite 102
Portsmouth, NH 03801

Book design by Tim Holtz
Printed in the United States of America

This book is dedicated to my brother, Rob Watson, who has consistently kept the flame of hope alive in the face of seeming hopelessness.

To my daughters, Samantha and Selma, for inspiring me to make a difference.

Contents

Preface	ix	Chapter 12	72
Acknowledgements	xv	Photo Section	75
Prologue	xvii	Chapter 13	81
Chapter 1	1	Chapter 14	92
Chapter 2	8	Chapter 15	95
Chapter 3	11	Chapter 16	105
Chapter 4	20	Chapter 17	110
Chapter 5	25	Chapter 18	112
Chapter 6	29	Chapter 19	119
Chapter 7	39	Chapter 20	123
Chapter 8	47	Chapter 21	136
Chapter 9	50	Chapter 22	146
Chapter 10	63	Epilogue	151
Chapter 11	66	About the Author	167

Preface

When I started writing this book, it was because I felt the stories were good stories. Most everyone likes a good story. Then, as I began digging into the stories, the theme of our collective humanity began to emerge. The daily activities at the dump are the stories of our lives, of our connection to community. And while there is always room for the humor, or the frivolous, there is an earnest undercurrent that happens everywhere—our lives produce waste, discards, leftovers and we're tasked with making decisions about this stuff very regularly, usually towards the lowest common denominator. We humans are faced with making upwards of thirty-five thousand decisions each day, and invariably this includes throwing things away. Originally, I had set the title of this book to be *Tales from the Dump*, but my daughters starting riffing new titles and from there *Everyone's Trash* was born. And the double entendre? Well, the title is meant to imply that everyone makes trash, not that people are trash, and part of managing our lives includes dealing with our waste. Because everyone makes trash, everyone has an opinion about it, even if you've never paused to think much about your opinion.

When I first purchased my home in southwest New Hampshire a quarter century ago, it didn't take me long to stumble on the dump that had long served that property. Bits of clay, some very rusted metal, and a clearly unintentionally discarded early 1800s penny were all buried down the hill from the house. The organic material that probably comprised most of what was discarded has long since been degraded and become part of the soil. As life sped up, the backyard dump morphed into the backyard burn barrel, and eventually the town dump became the place to consolidate the discards of life. And how complex it has become. Just today I received a petition asking me to sign up to make July plastic free. It's almost impossible to imagine how I could reasonably do that given how ubiquitous plastic has become for the modern lifestyle. It's not that it can't be done, but it would require a tremendous and thoughtful effort to succeed. And if the petition serves to expand our collective consciousness to think about how we live and how we use things and how we discard things, then it's a worthy effort. I remember precisely the day that my environmental ethic came rushing into my consciousness. It started with an act of senseless destruction.

As a pre-teen boy, I was very fortunate to be able to spend my summers with my grandparents in the Hamptons on Long Island, NY. I can still close my eyes and remember the days filled with fresh-picked blueberries mixed with Dannon blueberry yogurt, potato fields where I simply crossed the street to dig up a portion of that evening's dinner. Biking to the beach balancing an inflatable raft where I would hope for a yellow-flag day meaning it was "safe" to go into the water, but that the waves would be huge. Tuna fish sandwiches with the requisite sand stuck to the bread, the smell of my grandfather as he returned from New York City for the weekends—he was a mix of aftershave, scotch, and leather (from the old chair he sat in). My boyhood summers were Norman Rockwellish. And those times that the beach wasn't on the agenda or when my playmate brother was involved in other things, there was the small pond in the backyard. My very own Walden Pond. It was maybe ten feet across, but there was a cornucopia of life going on there and it utterly fascinated me. There was some open water, but lily pads dotted the surface, cattails ringed most of the shore, and insect life, particularly dragonflies, thrived. What most captured my attention though was the frog population. They could often be found sunning atop a lily pad and I would gaze in amazement watching their throats hum with their breathing and heartbeats. Sometimes

I would try to spur some action and I would aim a pebble at one of the frogs hoping to coerce them into jumping off the lily pad. Sometimes it worked, sometimes it didn't.

One summer day, a particularly stubborn frog did not react at all to my entreaties to jump, or do *something*. And it was too far into the water for me to wade in to catch it. Not far from the pond I found a rock, a pretty big rock. It took some effort to lift it over my head and I staggered back to the pond and heaved it towards the frog still perched on the lily pad. The effect was more dramatic than I expected. A veritable tsunami ensued and it took a very long time for the water to settle. And the frog was no longer on the lily pad. A few moments later it came floating up to the surface. Dead. I had not made the connection that this would be the result of my actions, but I had just actively killed a living creature for absolutely no reason. The emptiness that I felt in that moment, realizing that I had senselessly killed this beautiful frog, was profound. And yet that death kindled in me a fire that burns as bright today as it did when I realized that senseless killing was simply wrong. In killing that frog I killed a part of me as well, yet the act gave birth to my environmental ethic. I have what amounts to an almost irrational reverence for amphibians and the reptilian ancestor of the dinosaurs—turtles. I go out of my way to save frogs and turtles crossing the roads.

Such is my love of amphibians that when the Harris Center for Conservation Education, a local science and education organization that seeks to engage citizen scientists in conservation activities, contacted me about furthering their amphibian migration program, I leapt at the opportunity. In Keene, there are two primary locations where frogs and salamanders cross the road to get from their winter hibernation habitat to their breeding wetlands. The ideal conditions are darkness, plus 50 degrees F, and rain. The Harris Center terms the nights where the conditions are ideal for amphibian movement "Big Nights," and if you ever bear witness to this miracle of nature it's a sight you'll not soon forget. On a big Big Night, literally hundreds and hundreds of peepers, tree frogs, wood frogs, toads, spotted salamanders, and Jefferson salamanders all get the group email that it's time to move and move they do. As humans, we're conditioned to not dawdle whilst crossing the road, but the amphibians take their time, perhaps basking in the relative warmth of the road. This makes the amphibians particularly vulnerable to getting squashed by cars traveling through the area. Even one car moving through a Big Night zone can kill dozens of amphibians without even realizing it. The Harris Center trains volunteers to help usher the amphibians across the road, and when approached about a partnership that would close those sections of roads where migration is heavy, it seemed the minor inconvenience of being detoured around the migration area was a small price to pay for the many amphibians that would be saved from car tires.

Several years into this project and I type these words with the peace of knowing I'm participating in a good thing. Amphibians occupy an important part of the food chain and play a role in maintaining biodiversity through being predators for invertebrates and prey for owls, hawks, herons, turkeys, and foxes. Last year during a Big Night, I watched a barred owl take up residence in the migratory corridor and it feasted on any number amphibians before taking silent flight to digest. If you ever see things like this you don't soon forget.

The Big Nights aren't my only amphibian efforts. Warm, rainy spring nights during amphibian migration, I work by the headlights of my car shooing the frogs (and a lot of salamanders) across the road in various hot spots that I've come to know. I've saved hundreds of amphibians from being ignominiously flattened by cars. I also keep my eyes out for migrating turtles and I have a blanket and shovel in my car at all times in case I run across a big snapping turtle that is trying to cross the road. It's unbelievably satisfying to help an amphibian cross the road when the likely alternative is considered. And I'll never forget that frog in Long Island. My remorse is tempered by having learned a valuable lesson about the critters that occupy our shared environment. I can still feel that emptiness of that senseless day though.

It was pretty clear that while I excelled at biology in high school due to a fantastic teacher (thank you Mr. Hutchins), I just didn't like the lab work as much as I enjoyed systems, processes, commerce, and sociology. The love of the environment and frogs and turtles morphed into a more pragmatic exercise of my personal environmentalism. And while the tales from the dump are often the base of our human experience, the stories are instructive. We are dependent on each other, as we are dependent on the ecosystem that sustains our life. Our use of resources continues to expand, continues to deplete, and we adapt, we survive. Nature doesn't need our help; it does just fine on its own. The saying "Save the Earth" should really be "Save the Humans," because we're the ones despoiling the planet to our eventual existential expense. Finding our equilibrium within this harsh system is our quest. And there is no there in the "getting there." It's what happens along the journey. You are part of this story because one way or another you make up the stories in *Everyone's Trash*.

Acknowledgements

I've been blessed to work for a community that has a value system that largely mirrors my own. My colleagues at the Public Works Department in Keene, New Hampshire, are dedicated public servants, and the life of our bucolic city flows through the veins of the infrastructure we maintain every single hour, every single day, no matter the weather, no matter the situation.

We could never do the work we do without the support of our elected officials who represent the interests of the community at large. All of us who work at Public Works understand that economics and environmentalism are not mutually exclusive concepts. When we come before the city council, we know we have to make a business case while factoring in the environmental impacts to get the resources to support our initiatives and projects. I've often said that the best part of my job is its living laboratory aspect. Keene is big enough to have financial resources to advance public works and small enough that you can literally see the impacts of your work, practically in real time. That's pretty satisfying.

Prologue

Around the time I got my first paid job working at the Tiburon, California, recycling center, my life took a decidedly curious turn from a career standpoint. Of course, twelve years of age is not normally when careers are born, but a future that I couldn't have possibly anticipated dropped in my lap and temporarily waylaid me from my eventual career at the dump. I was attending Reed Elementary School and whatever day it was unfolded like every other day—recess punctuated by pesky classes. Recess was by far my favorite subject as it meant games of bombardment (also known as dodgeball) or pumpkin ball, which was baseball played with an orange ball the size of a volleyball you bounced into the plate and which had the added feature that you could peg a baserunner with the ball as another means to achieve an out. I didn't have many close friends, I just loved the competition.

On my way back to class from recess, I noticed a number of kids had formed a queue outside a classroom, not my classroom mind you, and without even thinking about it I joined the line. The line moved steadily until I found myself in a room with two adults with a reel-to-reel tape recorder in front of them. I sat down in the designated seat and they asked me what role I wanted to read for. It was then that the awareness set in that I had no idea why I had been standing in line, no idea what they were asking me, and no idea what was going on. I shrugged my shoulders, said I didn't know and they suggested, "Okay, why don't you try a few lines for Charlie Brown." Sounded good to me. They turned on the tape recorder, handed me a script, and asked me to read three lines of dialogue. Thirty seconds later they said thank you very much and I was on my way back to class. I didn't give what I had done a second thought, never told my parents what I had done as it seemed unremarkable. I did, however tell my parents about crushing the pumpkin ball over the centerfielder's head.

Weeks later, my parents received a call from Lee Mendelson Productions inquiring if I would be available to come to San Francisco to do another voice test for an upcoming Charlie Brown television special they were creating. It would take about twenty minutes and they were going to pay me fifty dollars for my "efforts." Woohoo! *Fifty dollars?* I was making something on the order of three dollars an hour at the Tiburon recycling center, so the answer to making fifty dollars was a resounding yes. A few days later, we jumped into my mom's blue Volvo station wagon and made the trip around the peninsula to Sausalito, across the Golden Gate Bridge, and into San Francisco where the recording studio was located. There was a small room with acoustic paneling on the walls and ceiling, a lone microphone in front of a stool, and a music stand with a script. Directly in front of the microphone was a glass-fronted control room loaded with toggle switches, knobs, buttons, and things that would simultaneously slide up or down with corresponding red and green lights. This was extremely cool.

I sat down in front of the microphone, put headphones over my ears as directed, and a voice came through the headphones asking me to read the first few lines of the script in front of me while being sure to speak directly into the microphone. As promised, twenty minutes of total time in the studio and my mom and I were on our way back to Tiburon where she dropped me and my bike off at school. Once again I forgot all about the audition. It seemed more trouble than it was worth to explain to my classmates where I'd been so I just didn't say anything. A month later my mom got a call from the production company saying they had narrowed down the field from the voice auditions and would I be willing to come in for another voice test the following Tuesday for which I would be paid another fifty dollars. Ummmm, yes! I was completely unaware of any competition associated with what I was doing because I was simply reading lines in my voice. My mom and I followed the same pattern as the first voice audition, and once again a significant amount of time went by before we heard from the production company a third time. They had narrowed down the candidates to be the next voice of Charlie Brown to six kids, me being one of them. I was asked to come in once more for a more extensive voice audition with the corresponding compensation upped to one hundred dollars.

While my drama was unfolding, my brother, a burgeoning Eagle Scout, constructed a shoeshine kit, with plans to launch his entrepreneurial career by plying shoeshines on the commuter ferry that ran several times each morning and evening from the dock in Tiburon to the Embarcadero in San Francisco. One dollar for a shine, and with the flair my brother developed, the snap of the polishing cloth and attention to detail, he developed quite a following and would finish his day pouring bills and coins on the carpet and counting his haul. On a good day, he would make forty dollars working from six to nine in the morning and again from four to six in the afternoon. The operators of the ferry were kind enough to allow him to ride for free, and his overhead was low. He was building up a healthy supply of cash which made me completely envious. I loved helping him stack the coins or arrange the bills, and every once in a while when he had something else going on he'd let me substitute for him, although my best hauls were about half his average take. I just didn't have a knack for snapping that polishing cloth and I was painfully shy about asking the suited men if they wanted a shine.

My aunt was visiting from the East Coast when the call came in. I had been selected as the voice of Charlie Brown for the upcoming *Peanuts* special "Be My Valentine, Charlie Brown." I believe my original contract was for nine hundred dollars, which pretty much felt like winning the lottery. We danced around the dining room, my mom, aunt and I celebrating something that didn't feel like it had much form or function, but the idea of nine hundred dollars was beyond my imagination. My dad must have been at work when the call came in, and my brother was definitely out shining shoes.

Over the next two years, I was the voice of Charlie Brown for four productions: *Be My Valentine, Charlie Brown*; *You're a Good Sport, Charlie Brown*; *Happy Anniversary, Charlie Brown* (the twenty-fifth anniversary special); and the feature-length movie *Race For Your Life, Charlie Brown*. In addition, I got some commercial work which required me to fly down to Hollywood to do the recording. I would get picked up at school in a limousine, driven to the airport, put on a plane, met at the airport by another limousine, taken to the recording studio, go back to the airport, and a limousine took me back to my house. The flight attendants (back then they were called stewardesses) were extremely solicitous towards an adorable twelve-year-old, earnest boy, and even though I felt far from it at the time, when I look back at pictures of me at that age, I was darn adorable, at least to adults. But on those trips, I still felt far from home, even though I had been raised to be pretty independent—navigating the streets of Mexico City at age eight with my ten-year-old brother by ourselves for two years when my father was transferred there with IBM, and, upon arrival in California, being given a bike and told, "If you want to go somewhere, you have your bike, don't ask us for a ride." That may seem harsh by today's 24/7-monitored children, but for me it was the ultimate freedom. Anywhere I wanted to go, all I had to do was peddle my way there. I took advantage of the friendly flight attendants and got more than my fair share of the meager offerings on the short flight from San Francisco to Los Angeles (chicken bouillon being my favorite treat).

Sometime in 1977, I was called in to San Francisco to begin the voice recording for *It's Arbor Day, Charlie Brown* when *it* happened. While reading one of my lines my voice cracked. I tried to read the line again and it sounded like I had a frog in my throat. Like the voices before me, puberty derailed my Charlie Brown train.

I was required to join the Screen Actors Guild when I signed on as the voice of Charlie Brown. I also got an agent who arranged other auditions for voice work and on-screen acting. I had zero acting experience and during the few things I auditioned for I was acutely aware of how self-conscious I felt in front of the camera. Aside from the Charlie Brown work, nothing else was coming my way, and now my voice-acting career was over. I did go on one final audition for a movie, and I was away for a week hiking with the Boy Scouts (unlike my Eagle Scout brother, I was in the Boy Scouts for two years and didn't earn a single merit badge) when a call came in about casting me in a lead role. Because I was away, they went with the second choice and the veritable crossroads of one career versus another placed me on the trail to the dump. The movie turned into a huge box office success. Who knows what might have been. The next weekend, I was at my attendant job at the Tiburon Recycling Center, an unknown harbinger of things to come.

I made enough money through my voice acting to eventually pay for my freshman year of college. There is something very Charlie Brownish about reaching your peak earning at twelve years old. It's been downhill ever since. Most voice actors inhabit characters, but not so with the *Peanuts* gang. The voicing was done by children because they wanted the characters to sound like kids. I was fortunate that my voice acting didn't require any particular talent. My voice had the right sound, and I was lucky to have been in the right place at the right time. I was also lucky to work with Bill Melendez, one of the producers of the Charlie Brown specials who did the voices for Snoopy and Woodstock. He was a crazy man who inspired me with his child-like wonder for the work he did. I still occasionally get a residual check for work I did forty-seven years ago. It's like finding money on the ground. And while my children think it's kind of cool that their father was the voice of Charlie Brown, I hope that my legacy of the work that I've been doing for the past thirty years is ultimately more noteworthy. And if you must know, the voice of Ms. Othmar, the teacher in the *Peanuts* shows is a trombone, the voice of Peppermint Patty during my time was a boy, and Charlie Brown grew up to be a hopeful environmentalist just like my Eagle Scout brother. Good grief.

It should be noted that my mother attended a parent-teacher conference when my brother was in kindergarten where the teacher told my mother, "Mrs. Watson, I've been a teacher for twenty-five years, and I know a criminal when I see one." In addition to his Eagle Scout achievement, he managed to channel his criminal mind towards an undergraduate degree from Dartmouth College, a master's degree in systems design from U.C. Berkeley, and an MBA from Columbia University. He also founded the Leadership and Environmental Design (LEED) green building rating system, founded the Solid Waste Environmental Excellence Performance (SWEEP)—the only national voluntary sustainability standard for the solid waste industry—and was named one of the twenty-five most influential alumni in Dartmouth College history. I think it's safe to say that Ms. Wick, may she rest in peace, got that one wrong.

Chapter 1

It wasn't like I aspired to work at the dump. My motivation, at first, was very simple. "You're going to pay me to break glass?" This was a major job perk for a twelve-year-old. "When can I start?" And so, the following Saturday found me as the official attendant at the Tiburon, California, recycling center located directly across the bay from San Francisco. The year was 1976, and even "the dump" was caught up in bicentennial fever with some mangled bunting adorning the various dumpsters. The gravel lot of about a half-acre was fenced in, from what I cannot imagine, but when the gates were open there was an unencumbered view of the San Francisco Bay, the skyline of the city, and of course the iconic Golden Gate bridge. It's easy to look back on the scene forty years later and recognize it for the aesthetic paradise it was, but for a young boy it was about getting paid to break glass. I was the lone attendant on the only day the dump was open, from eight-ish in the morning to about 1 PM. It was never a bustling place, but a steady stream of customers would arrive over the course of the day. Glass, steel cans, aluminum cans, and bundled newspapers which, I learned, told me which newspaper was winning the newspaper war: the *San Francisco Chronicle* (arguably had the better sports section), or the *San Francisco Examiner* (better comic section), with the lowly *Marin County Independent Journal* always a sad afterthought with neither good sports nor comics. The customers would place the various items in front of the dumpster designated for that particular item, except for the newspaper which was tossed into a forty-cubic-yard roll-off container. Unbeknownst to me, my job was ostensibly quality control. Making sure that the clear glass dumpster only had clear glass, I would do my best John Montefusco impersonation and go through my windup.[1] Runner on first base, two outs, count three balls, two strikes. Montefusco goes into the windup, it's a heater, swing and a miss strike three as the satisfying explosion of glass against the steel wall of the dumpster accompanied each and every pitch. Let me repeat, I was getting paid for this!

The job took an unexpectedly serendipitous turn when one day while inspecting the newspaper roll off I spied a *Playboy* magazine perched on the very top of the pile of thousands of pounds of paper. It was as if I'd discovered the Holy Grail. I'm certain that the official term "dumpster diving" was coined much later than 1976, but my arc of descent from the platform into that dumpster was Olympic worthy. I landed on my prize, climbed out of the dumpster, and surreptitiously tucked the *Playboy* into my backpack for later examination. Then I began to wonder if there were more of these prizes. The glass breaking didn't seem quite as fascinating as the mine engineering required to dig into a full dumpster in search of the elusive golden nuggets, because quite frankly, at twelve years old, these things were worth their weight in gold. Turns out that with the diligence only a young boy in search of seeing a naked woman can muster, and with some rudimentary engineering skills, tunnels of questionable stability could be constructed, and on any given Saturday I'd be the proud owner of half a dozen magazines that were saved from recycling to live one step above in the hierarchy of environmental stewardship: REUSE!

I was vaguely aware that this job made sense. I knew that the stuff I was helping to sort was going to be made into new stuff, and that seemed like a logical thing to do. But my introduction to the world of the dump was this half-acre plot, and what happened to the material when it left

---

[1] John Montefusco was a pitcher for the San Francisco Giants from 1974 to 1980. He was dubbed "The Count of Montefusco" by then play-by-play announcer Al Michaels. I used to sit on the rocks on the breakwater of the San Francisco Bay listening to Al Michaels call the Giants games. It was heavenly.

was a mystery. I worked at the Tiburon recycling center until my family moved to New Hampshire before my sophomore year in high school. There was no recycling center in my new hometown, and my only connection to trash was my father's proud addition to the arsenal of helpful kitchen appliances—the trash compactor. This was not much more than a metal, kitchen trash bin that when closed and activated would engage a hydraulic ram that squished the trash just a little better than if you put your foot in the bin and pressed down. It was somewhat satisfying to hear the glass bottles being crushed, although, the inevitable torn trash can liner was a definite flaw in the system. I was mainly caught up in my high school days and recycling was an afterthought for a number of years. It was when I was looking to change careers in my late twenties that my interest in recycling was rekindled.

My college degree was in hotel management, a logical choice due to having been in my family's restaurant business starting at age fifteen, and I took a job after graduation with one of the big chains. A couple years of that soulless work and I knew I had to do something different. There's almost no better way to completely change a career track than going back to graduate school. I landed at Antioch University–New England in Keene, New Hampshire, after hotel stints in San Antonio and Dallas, Texas, a brief period managing my family's restaurant back in New Hampshire, and a year finding myself as a ski bum in Park City, Utah. Antioch was a great fit. At this crunchy school with a solid pedigree in environmental studies, I decided that I was going to take my business degree and use it to learn the business of recycling. Unbeknownst to me, the wheels of destiny were turning that would affect my life and the lives of others.

Seven miles away from the sheltered confines of Antioch University is the Keene, New Hampshire, recycling center. In 1989, with the most recent wave of environmental awareness that was sweeping the nation, the city of Keene had restarted a recycling program that had been dormant since the scrap drives of World War II. It was a crude operation that was housed in a two-car garage adjacent to an unlined landfill. The workers included Ronald D and Michael S. The job they performed included sorting the recyclables collected at curbside by private haulers and baling them to be sold as commodities. The city offered free recycling versus a charge to put stuff in the dump as an incentive to get people to recycle. The job at the recycling center was labor intensive in a place that didn't smell very good and required hearing protection to drown out the noise of the equipment and the constant sounds of glass, steel, and plastic churning their way to eventual new life. On any given day, there were six to nine workers on site, including the guy operating the compactor on the landfill, the gate attendant who always answered the phone with a less than cheerful greeting of "DUMP," three to five guys sorting recycling (depending on how many welfare-to-work workers were dropped off at the dump), and a manager. There was as much of a hierarchy to the organization as there was efficiency in the operation. In other words, not much.

Ron and Michael were two of the six paid staff. By benefit of longevity, Michael was Ron's supervisor. Ron and Michael had some history together. In a twist of fate, Ron and Michael had been at the Cheshire County Corrections Facility together. Ron as a prison guard, Michael as a prisoner. Michael had been in trouble with the law since his teens and was serving a sentence for his role in an armed robbery when Ron was a guard at the prison. Previously, Michael had been convicted of a knife attack. Now he was Ron's supervisor. Bullying was a much more common practice in those days and Michael had honed these skills to an art form. His preferred target was Ron. Ron was married, had two young children, and served in the National Guard. Day after day, Michael made Ron's life a living hell. When the normal supervisor to subordinate crappy work assignments failed to satisfy, Michael began a systematic psychological torture, probing Ron for weaknesses. During work breaks, Michael would often brandish a knife, slowly sharpening it on a stone and making comments about the various ways he might use the knife directly or indirectly to harm Ron. The harassment continued for months. Ron reported the harassment to the manager, he reported the harassment to the police. The response was always the same: Michael hadn't done anything and it was only words. Michael saw an opportunity to ramp up the torture when he learned that the following week Ron was due to deploy out of state for two weeks of National Guard training. Threatening gestures were nothing new for Michael, but his new mode of attack ramped up the level of evil. On a sunny morning two days before Ron's deployment, all the workers were on a lunch break, Michael sitting opposite Ron in chairs that had been salvaged from the landfill. Michael, as he often did, was sharpening his knife when he looked up at Ron and said in a voice loud enough for everyone, including Ron to hear, "When you go off to your National Guard training, I'm going to go over to your house. When I get there, I'm going to kill your children, rape your wife, and burn your house down." Ron didn't say a word. He got up from the group and while walking to his truck heard Michael cackling. Ron drove off. His house was only a few minutes away. Michael must have felt jubilant that he had affected Ron so profoundly, but a few minutes later Ron drove back into the recycling center parking lot. As the dust settled around his truck, he opened the door and stepped out brandishing a 9mm pistol. He located Michael and strode towards him. Michael saw Ron had a gun and turned to flee. Two bullets tore into his back as he fell to the ground dead. Ron calmly walked over to Michael, stood astride him, and then emptied the reminder of this clip into Michael's lifeless body. Thirteen more bullets. Ron then sat down, placed the gun in his lap and waited for the police to arrive.

Ron was arrested and charged with first degree murder punishable by life in prison. The initial investigation did not indicate any premeditation by Ron. This was more of a case of a man being brought to the breaking point and beyond. A videotape of Ron in the interrogation room showed him praying for atonement as well as for the victim. Coworkers were interviewed, the manager was interviewed. Story after story revealed Michael to be an unusually cruel and intimidating bully. Had Michael been capable of carrying out the rape, murder, and arson threat? The grand jury thought so, and it declined to indict Ron. Ron was a free man, but the man in charge of all city employees, the city manager, had a dilemma on his hands. There was no disputing that Ron had killed Michael, but there was no provision in the city's personnel manual about the discipline appropriate for killing a fellow employee. Ron was terminated under some vague provision of conduct unbecoming of a city employee. Eventually Ron moved out of the area. The police file, for some unknown reason, remains open and in order to look into it further I would need to contact the state attorney general's office. This tragedy set in motion the wheels of my reign at the dump. It was decided that the murder was due, in part, to gross negligence by the manager, who should have known of the tension and the possibility of escalation. It was a scapegoat setup, and the manager was fired. The city of Keene began to look for a new manager of the dump, or in more polite terms, a "solid waste coordinator."

Weeks away from graduating with a master's degree in resource management, I was walking down the hall after class and I happened to stop by the bulletin board to see if there were any job postings. There was one, and it was for something I just happened to be qualified for. Two years of coursework focused on managing waste, a voluntary internship at the EPA, a practicum at the Environmental Hazards Management Institute, and a stint as the administrative director of the Massachusetts Recycling Coalition had led me to this destiny. There were fifty-seven fellow applicants for the solid waste coordinator job. In my job interview, the public works director posited a hypothetical question: "If you have twenty tons of green glass, but it costs you $50 a ton to recycle it or you can bury it in the dump for free what would you do?" I wasn't exactly prepared for this type of question, but I responded by saying that I would find it difficult to believe that those were my only two choices. Given those two choices,

I would bury the glass, as it is an inert material and there wasn't going to be a shortage of sand and silica (the raw materials to make glass) anytime soon. In my mind the business of recycling had to make sense from an economic standpoint as well as environmentally. To me these were not mutually exclusive concepts (more on this later). Apparently, this was the right answer, as the public works director stood up from behind his desk, extended his hand, and offered me the job. With more than a little seriousness, the handshake lingering, he said, "And the measure of your success will be to make sure no one gets killed at the dump." Thirty-two years later, so far so good.

Chapter 2

Changing careers isn't unusual in our modern economy, and going to graduate school felt like my ticket to do something professionally with a soul. My sense was to take advantage of the doors that would open during graduate school to expose myself to the business of recycling. Knowing I needed to shift my resume from the hospitality world to the recycling world, I met someone from the EPA at a conference and talked my way into a voluntary internship at the EPA Region I office in Boston. I spent a semester traveling by bus from where I lived on the North Shore of Massachusetts into Boston two days a week. I was assigned to the Waste and Recycling Division and given a cubicle along with a vague assignment that involved researching how to increase green glass recycling.

Green glass recycling turned out to be a curious problem. Glass is among the most easily recyclable materials, and a glass bottle can be melted and turned into a new glass bottle an infinite number of times. The problem stems from the fact that most beverages produced domestically are bottled in clear or brown glass. At the time, only Rolling Rock beer used green glass in domestic production, so while lots of imported beer was bottled in green glass, it was not economical to ship the glass to be recycled overseas, and Rolling Rock had all the green glass they could ever want. The end result was a product that had no marketable value that was being recycled in significant quantities. As with all sustainable recycling processing, the economic resource part of the equation has to work, and for green glass it was clearly not working. My job was to see if there were ways to improve the recyclable prospects of green glass. Believe me, many a great mind has pondered the green glass questions before and come up empty, and I wish I could report that I discovered some brilliant solution, but mostly I just confirmed the original problem. There was, however, an interesting interpersonal dynamic that began to unfold between myself and one of the women in the secretarial pool.

I'm sure that the administrative professionals at the EPA in the office where I was assigned had much better things to do than support some snot-nosed intern who might occasionally need a file or to get a phone message, and as a general rule I'm unfailingly polite, especially to administrative people because they can make or break you. The hostility of this one particular secretary took me off guard. As I was only at the EPA two days a week, there would be days when I'd get returned phone calls which the secretaries noted on the pink "While You Were Out" notepads. Yes, this was at a time when e-mail was just beginning to appear, administrative assistants were still called secretaries, and I was using a Macintosh SE 20 featuring a twenty-megabyte hard drive for a bit of context. So phone messages were a primary form of being aware of missed calls. My second week at the EPA, I was handed a small stack of messages with my name prominently printed on the top of the message "DUNKIN." As she handed me my messages, of course, I noticed the misspelling of my name and politely pointed out that my name is, in fact, spelled D U N C A N. She looked at me blankly, her eyes never blinking, and then resumed her far more important task, which in that case consisted of painting her nails. This pattern repeated itself for several weeks—a stack of messages, "DUNKIN" printed at the top, my gentle reminder that it was still Duncan. The sixth week of this game and she finally spoke to me. When I again corrected her on the spelling of my name she uttered her first words to me, "What difference does it make?" Like the bull who sees the red cape and charges in anger I completely snapped. Arms flailing I flapped in circles saying out loud for everyone to hear, "What difference does it make!? What difference does it make!?" I circled back to her, put my hands on her desk, leaned over looking her in the eyes and hissed, "I'll tell you what difference it makes! One is a fucking donut shop, and one is my name, so how about from now on I call you shithead because what difference does it make?" I stormed back to my cubicle seething, and other than slightly widened eyes I got no satisfactory reaction from her. But the next week when I went to retrieve my messages they were all labeled at the top for "Duncan."

As for the green glass, the situation decades later is not much improved as far as recycling. In Keene, we either crush some of the glass to three-eighths of an inch minus specification and use it as a construction aggregate for the base of roads and sidewalks, or we sell it to a company that uses it to make fiberglass insulation. But let's face it, while so many products have shifted from glass to plastic, a beer never tastes as good as when it's bottled in glass. Maybe some of the burgeoning craft beer brewers can help out by sourcing green glass to bottle their product (but most craft beer I know is packaged in aluminum, so I won't quibble with the recyclability of their product).

Chapter 3

I don't look like a crunchy tree hugger. The bushy mane of my younger days has been replaced by a lot less hair that I keep high and tight. While books should certainly not be judged by their cover, I have found that my clean-cut first impression allows me entry into doors that might be closed to alternative thinking. But there are many, many people who simply love dump culture. And not everyone is stereotypically grouchy as depicted in some of the popular culture dump dwellers—Fred Sanford of *Sanford and Son* and Oscar the Grouch who literally lives in a trash can are two iconic examples. I know plenty of people who might not freely admit to it, but when passing by a dumpster are known to step up for a look to see if maybe there might be a treasure or two hidden there. And if you haven't ever done this, it's because it's a well-kept secret by those who have discovered occasional jewels inside. Try it sometime. I'm not advocating full-on dumpster diving, as I don't want to encourage people to go rooting around a dumpster in an unsafe manner (and frankly there aren't any safe manners for rooting around) but take a peek sometime. You just might be surprised by what you see.

My personal dump affliction is to notice all things associated with recycling or what some might more directly call trash. Such was the case when my then spouse and I spent a year in Utah being ski bums. I arrived in Utah after I had spent a year running my family's restaurant after the death of my mother. I guess no one expects to lose a parent to suicide and I didn't have any inkling that things were serious, but many years before—after my birth—my mother came down with a severe case of post-partum depression and had tried to kill herself after receiving some terrible therapy from a Freudian therapist who indicated that my mother's depression stemmed from some truly sick and twisted idea of a (non-existent), sexual connection between my mother and her father. The very idea of a twenty-year-old mother with two children under the age of two being told by an uninformed fabulist of a therapist to believe in a non-existent sexual assault she'd somehow suppressed the memory of being the source of her anxiety is so beyond the pale of responsible psychiatry that it still is difficult to process. But the first suicide attempt failed, my mother healed. She was an amazing mother and used her trauma from therapy to become a therapist herself. I met many people who came to her memorial service, dozens and dozens of people who sought me out to tell me how much my mother helped them through terrible ordeals. As I've come to understand it, a perfect storm confluence led to my mother's rapid demise. She broke a cardinal rule of therapy with a particularly difficult problem and allowed the wall that must exist to protect the therapist from taking on the emotional turmoil of their patient to come down. Then she got the flu. A really bad flu, and if you've ever had the full-blown flu as an adult, then you understand it is debilitating. I think it's safe to say that when you have a high fever you aren't exactly in your right mind. The difficult patient, the flu, and my mother possibly being bi-polar led to a deep depression from which she decided the only way out of the pain was to end her life. In a peaceful cemetery a few miles from downtown Hanover, NH, my mother ran a hose from the tailpipe of the Volvo station wagon into a cracked window, got back into the car, turned on the ignition, and died of carbon monoxide poisoning. She was forty-eight years old. While I know this doesn't have anything to do with the dump, I just have to share this one bit about her death that makes me believe there is a lot more to our lives after we die. And that feels hopeful to me.

    My father called me at work. I don't remember what day it was, but I was at my desk as a sales manager for the Hyatt Regency at the Dallas-Fort

Worth airport. My father said that my mother was missing and he feared the worst. My mother had left a note that seemed to indicate that she was spiraling into a void that threatened self-harm. After my father had combed all the usual places my mother might go, he called me and my brother in a panic. I caught a flight later that day and within hours my brother, having flown in from California, met us at our home in New Hampshire. It's a daunting prospect to think of where to look for someone who has gone missing, particularly in 1988 with no cell phones, no social media, only the corded telephone to communicate long distance. It was late April, the day was reasonably long, and as we ran out of daylight without finding any trace of her, we hunkered down for the night hoping she might call to check in. I was restless, but I needed sleep and I settled into my old bedroom around 11 PM ready to resume the search at daybreak. Around 1 AM, my then wife and I were suddenly roused by a blindingly bright orb that had entered the room. We both sat bolt upright as this spit of energy hovered, crackling with energy, bright as the sun. The orb hung, bouncing a half inch in every direction simultaneously, ten seconds, maybe twenty, then at the speed of light it shot out the window and it was gone. I looked at my then spouse, confirming that we had indeed seen the same thing, then somehow managed to fall back to sleep.

 I was awakened in the morning to the sounds of crying coming from the kitchen. I threw on some pajamas and made a beeline to the kitchen where my father and brother were in an embrace, sobbing. A police officer had just departed after telling them that they had discovered my mother's body at the cemetery. I joined the sobbing embrace and realized that my life had just changed in ways that I could not possibly imagine. I later found out that the time of death coincided precisely with the timing of the visit by the orb of energy and light. It's hard to part ways with those you love, particularly when they go before what seems like their time. And if there is one thing that is inexorably true, it's that life goes on. Relentlessly.

In short order, I left my job working for Hyatt Hotels, assumed the day-to-day management of the family restaurant (for those who have ever been to Hanover, New Hampshire, the restaurant was Lou's, established in 1949, and the Watson family was the second owner, buying it from "Lou" Bressett), an iconic diner-style restaurant on Main Street. I spent the better part of the year with my nose firmly to the grindstone trying to keep a forty-eight-seat restaurant open twenty-four hours a day with almost seventy employees afloat during a time period of less than 2 percent unemployment in the area. It was a tough year.

As my father healed and was ready to come back to running the business full-time, I was thinking about graduate school but not sure what direction I wanted to go when a friend suggested going out to Utah to be a ski bum for a year. My friend assured me that my background would easily enable me to get a food service job in the evenings, leaving me free to ski all day. And before I knew it, I loaded up my Saab 900 with my then wife, my ten-month-old golden retriever (named Hyatt Regency), and all our worldly belongings. We had rented a renovated garage turned into a one-bedroom apartment close to town and spent a week driving to Utah. It was early September, 1989. Park City, Utah, was a jewel of a town. Long before the Olympics had even thought of coming to town, Park City was at the beginning of going upscale, but there was still a prominent funk factor as many of the locals who created the funk over the past several decades were still hanging on. But as Deer Valley started to take off and the stupid money began to discover the most amazing thing about winter in Utah—the light, fluffy powder snow that is so dry that it's nearly impossible to form into a snowball, Park City was full of opportunity for the ski bum lifestyle. There's a saying in town that people who settle in Park City come for the winter, but stay for the spring, summer, and fall. Arriving there in early fall, it was easy to see why. The base of Park City is at seven thousand feet. The towering peaks on the south end of town are not bordered by tall peaks to the west or north, which gives Park City a more wide-open feel, less claustrophobic than many of the Colorado ski towns I've been to. I loved the feel of the place, but quickly discovered that though jobs were plentiful, most didn't start until late December when ski season kicked into full gear. This meant I had a lot of leisure time the first few months.[2]

It did not take me long to discover that Park City had a magnificent recycling center. It looked like a place that had been designed by the hippies that danced in the musical *Hair* while singing "Aquarius." I loved it, and the variety of material they accepted was impressive even by today's standards. I dutifully began to collect material in our garage/apartment home. This was much more garage than apartment, as the guy who renovated it obviously had never heard of a level, nor thought to add heat. Snow that blew under the door during storms would remain there, unmelted, until I swept it out into the driveway. Every week I would make a dump run, talk with the long-haired attendants and then I was off to hike, bike, golf, or ski.

Early in the winter, I noticed a flyer posted at the dump. The first annual "Dash for Trash" cross-country ski race was to be held the following Saturday. I didn't like downhill skiing on Saturdays anyway because it was way more crowded than weekdays, and I had been on the cross-country ski team in high school. Why, I would dust off my old skis and participate in this worthwhile community event to benefit Park City Recycling Center. It seemed like a lark.

I arrived bright and early that Saturday morning ready to relive my glory days of skiing. After all, I had finished in the top fifteen in the New

---

[2] I had the opportunity to revisit Park City in the fall of 2022 on my way back from a backpacking trip in the Wind River Range in Wyoming. Park City today compared to 1989 is simply unrecognizable. The development has gone amok, consuming what used to be significant open space, as people flock there to try to capture some of the magic that I clearly saw in 1989, but it's now just an over-commercialized ski town catering to the rich and ultra-rich. It's pretty sad really. Joni was right—pave paradise and put up a parking lot (good luck finding parking in Park City, though).

Hampshire high school cross-country ski championships a mere eight years prior. I was young and fit. This would be a piece of cake. So what if I hadn't even put on cross-country skis in the past six years? I was full of hubris as I removed my sweat pants and jacket to reveal my skin tight ski suit which still fit. The tailgate of the Saab was open, my skis propped on the bumper, and I was taking out my sad toolbox filled with ski wax when two vans pulled into the spaces next to me. One was full of the University of Utah cross-country ski team, the other was filled with the US Olympic cross-country ski team. Then more cars pulled up with very, very fit, serious-looking people, and it was then I realized I was way over my head. I didn't blink, I carried on with my task of trying to figure out what I would use for wax.

Cross-country skiing has evolved somewhat since 1989. Now there are two styles: classic and freestyle. Freestyle uses shorter skis and you propel by skating with the skis in the same motion as ice skating. Classic style consists of two tracks set in the snow, and generally the skier stays in those tracks for the duration of the course. All cross-country skis have a camber, which is an arch in the middle of the ski. When weight is put on the ski, the middle presses into the snow and either a pattern or wax on the camber portion of the ski when pressed down grips the snow, allowing the ski to use that grip to propel the non-weighted ski forward.

As this race was a classic-style race, the skis needed to be waxed. Now, you might think that all wax is created equally, but nothing could be farther from the truth. Specific waxes work in specific conditions and there are literally experts that factor in every conceivable variable from temperature to snow type as well as how the conditions may change over the course, for example, will the skier be in the sun, shade, or both, etc. A wax expert I have never been, so I began to watch the more skilled skiers next to me for a hint at what wax to use, since it had been so long since I had needed to seriously wax my skis. Mind you, I thought I was going to be tooling around on a golf course with a bunch of recycling-loving granolas, raising a bit of money to keep the lights on at the recycling center. What amounted to a World Cup qualifying race is definitely not what I bargained for. Yet I persisted, going through the normal pre-race prep, waxing, stretching, and psyching myself up. What I failed to do was test my wax. I was in good shape, but not high-level race shape, and I figured I would conserve my energy.

I thought back to my last race in high school, which capped a rather undignified end of my racing career. I had arrived in New Hampshire to start my sophomore year in high school, and as it started snowing around Thanksgiving, a winter sport was a necessity. The established hockey community wouldn't accommodate a neophyte, I was truly terrible at basketball, but the ski team welcomed all. I admit that when it comes to sports I'm hyper competitive, and I love exceeding people's expectations of me. I had never, ever strapped on cross country skis, but I knew how to train, so I used the dry-land training to whip myself into peak shape. By the time snow covered the courses and the races started, I was placing and scoring points for my team, even though I was short on style and technique. I would never be mistaken for a natural cross-country skier, mostly because I was impatient to the necessity of having the unweighted ski glide forward. My skiing looked a lot more like running with long sticks attached to my feet, but surprisingly it worked. My particular strength, running up hills where everyone else was using a herring bone stepping system was very effective, and the people I whizzed by seemed demoralized, which delighted me and always propelled me faster. I was a middle of the pack guy to start, and I was never a threat to the top ten skiers because they had the fitness and technique to put them in a league of their own. I was simply wildly exceeding expectations, attracting the attention of the coaches, and I had a hunger to succeed.

I finished in the top fifteen in the New Hampshire state championships my sophomore and junior years and was named captain of the cross-country ski team for my senior year. By the time my senior year ski season came around, other priorities, like partying, my girlfriend, my job, and frankly everything else, seemed more appealing than dry-land training or racing, so I mostly went AWOL. I skied in a handful of races, did quite poorly, and I didn't much care. At the end of the year, I received a humiliating condemnation at the end-of-the-year banquet. I'm quite certain that this "award" has never been duplicated, but the cross-country ski team coach gave a long speech about what a disappointment I'd been as captain, then made a great show of handing the captain of the varsity ski team a junior varsity letter. It was kind of a jerk move, but I also probably deserved it.

It was with that tainted backdrop marking the end of my ski-racing career that I was now strapping on skis once again. My initial bluster and confidence was waning by the second. There were over one hundred people beginning to assemble at the starting line. Some in team uniforms, all but one sporting the skintight suit that supposedly makes you more aerodynamic, but until you really get going just serves to make you cold. The one exception to this intense race-ready crowd was a woman who clearly read the "race" flyer the same way I did. She had old-school, wide fish-scale skis, no wax required, and she had a backpack strapped to her back with a six-month-old cherub sticking out the top. That baby was far more content than me when the gun fired signaling the beginning of the race. This was for recycling dammit! There are few moments in sport that equal the hysteria of a mass start cross-country ski race. Every racer bursts off the line trying to be the first to get into the tracks and set the pace. I pressed my waxed right ski down to the snow to join the scramble, being just a tad late when everyone else leapt at the gun, and where I expected my ski to grip and propel me, the ski acted as though there was no wax at all and I slipped and fell on my face.

I got up quickly to discover that everyone but me was successfully on the course, including the woman with the baby. With the right wax, you grip and glide, grip and glide. With the wrong wax, you either grip and grip, or slip and slip. My wax selection was the slip and slip variety of a big screw up. A smarter person would have just called it a day. Technical difficulties. But I'm apparently a masochist and the only way to propel myself forward was to double pole. Double poling is a standard technique in classic cross-country skiing, but it's a circumstantial technique. It is not meant to be used exclusively for a ten-kilometer race—two five-kilometer loops over a hilly golf course. By the time I finished the first loop, a large contingent of racers, in fact almost all of them, had passed me and already finished the race. Again, discretion would have been the better part of valor, but I pressed onward. My arms had long given out. Each thrust of my arms up and forward was met with no strength as my poles planted in the snow. Foot by miserable foot the agony multiplied. I was sweating in a profound way. My race suit was a darker shade of red because it was saturated, and the sweat filled my eyes blinding through my frantic blinking.

I was having an out of body experience when I collapsed at the finish line. The attendants monitoring the time had packed up, and were milling about in case they needed to summon the emergency services for me. I ended up calling in sick to work for the next three days, and in hindsight I should have gone to the hospital to have fluids replaced. Somehow I managed to beat the woman with the baby, although I don't remember passing her. She came in a couple minutes after me, the baby happily chortling while I was splayed on the snow. I raised $20 for the Park City Recycling Center, so all in all it was a worthwhile endeavor.

Chapter 4

I've been known to be a person who, when I get a bee in my bonnet, can be somewhat insufferable. Such was the case when, in graduate school, living near the Seacoast of New Hampshire, I stumbled on an idea for a recycling experiment: How much recyclables do two people generate in a year?

The apartment we rented was essentially a house that had been sliced down the middle. Our end featured this sad little sun porch that had a depressing light and would never be used for more than a passageway to get to the driveway where our cars were parked. Our landlady, an elderly Polish woman with one of those impossible to spell last names and her German shepherd named Hershey, gave us the nickel tour. It was meager, but meager is de rigueur for two people trying to make ends meet while simultaneously in graduate school. With the signing of a one-year lease, my experiment began.

Since the town we lived in had no recycling center and the sunroom had no real purpose, I thought of it as the perfect place for a repository of accumulated recyclables within a calendar year. I placed the first bottle of what would become hundreds that first afternoon after schlepping our belongings from the moving van into the house that warm day on the first of June 1990.

It felt important to me to walk my talk. I would be starting school in the fall, and by God, I was going to save the planet all by myself. And frankly, I was curious. These things I would be setting aside because they were easily recyclable in a number of communities, including many surrounding communities, were what most people in my town would be throwing away. I think it was the statistic about the energy loss by not recycling an aluminum can that really got my attention. Someone, somewhere, did an energy analysis of the energy required to create a new aluminum can made from recycled aluminum versus producing that same can from mining virgin bauxite ore. At the time, the energy loss of throwing the aluminum can in the trash was equivalent to filling it three-quarters full of gasoline and pouring it onto the ground. Now the good news is that the majority of aluminum cans produced are recycled and aluminum recycling has been around since the inception of the aluminum can. But there are also millions of aluminum cans thrown away every week. The Aluminum Association estimates over a billion dollars of aluminum is thrown away every year, and the volume represents enough aluminum to rebuild the entire US commercial air fleet every three months. It's a stupendous waste, and I cannot help but think of millions of cans three-quarters full of gas being emptied on the ground. This haunts me, and whenever I see an aluminum can in the trash I'm not above fishing it out and putting it in a recycling bin when I find one. I don't do this to be smug, I just think of the gas.

Over the ensuing weeks and months the sunroom became increasingly laden with the standard fare of recyclables separated by type. I had bags for clear, high-density polyethylene plastic (HDPE), known by its symbol #2 on the bottom of the container. This type of plastic is commonly associated with milk jugs. Several years later, a prominent dairy introduced an opaque plastic milk jug that was purported to block light, thereby preserving the nutritional value of milk for a longer period of time. Skeptical of what I believed to be a marketing ploy, I had one of my colleagues at the recycling center close me inside a refrigerator to prove what I suspected to be true. The light does, in fact, go out when the refrigerator door is closed. So what if the opaque bottle sent recycling markets into a tizzy? The marketers had done their job, and opaque milk jugs are still produced. Which brings me to the next sort. I had a bag for colored, high-density polyethylene, which is typically associated with detergent bottles. Next was a bag of #1 plastic, polyethylene terephthalate (PETE), also known as soda bottles, a bag for aluminum cans, steel cans, a bag for glass bottles—okay—beer bottles, and a bin for mixed paper as well as a bin for newspaper. When a bag or bin was full, I moved it into the corner so as not to present an obstacle should we have to evacuate the house because of a fire.

About halfway through the year, I was asked what I intended to do with what had become a small mountain, dominated by newspaper because I had a subscription to the *Boston Globe*. The truth was I didn't know. I mean, I knew I wanted to weigh the material, but I wasn't quite sure where I was going to get rid of it.

A couple days a week I drove to the bus station and boarded the bus to go to Boston, where I toiled as the administrative director of the Massachusetts Recycling Coalition. It was a great job in the sense that I was making connections in the world I wanted to work in, but it certainly was not paying the bills. I went to school all day on Friday and most weekends, so I had another two days I could work. One of my undergraduate fraternity brothers lived in a nearby town and his job was selling golf equipment, including the premier line of golf balls in the sport. His territory was Maine, New Hampshire, and Vermont. Even if he had been extremely organized, it would have been a difficult territory to service properly, but he wasn't organized and he desperately needed my help. My job was to deliver product to his third-tier accounts so he could focus on the top-tier clients that represented 90 percent of his business. It was the perfect, mostly mindless, job to complement the mental energy I spent working on my master's degree. I drove all over northern New England, arriving at the assigned golf course, talking to the golf pro, finding out what they needed, and at my last stop the pro would always offer up a quick nine holes if I wanted. I always had my sticks in the car, and it was hard to resist playing the largely empty courses in the backwaters of Maine, New Hampshire, and Vermont.

As Memorial Day approached, marking a time to move to a new place, we had been offered a caretaking position for an old farmhouse located about thirty minutes south. The sunroom was now full from floor to ceiling with recyclables, resources that deserved a second life. I just needed to find a place to recycle this collection. The answer came in the form of my friend who employed me selling golf products. He had a van and he lived in a town with curbside recycling. It took three trips with his van and my own car loaded to the gills to transport what turned out to be 970 pounds of recyclables. I know, because I picked up each bag and each bin, stepped on the bathroom scale, weighed myself, then did the math to subtract my body weight from the reading on the scale while holding the bag or bin. The Thursday before Memorial Day weekend, my friend's garage was filled with my recyclable discards ready to be placed at curbside for the scheduled Friday collection.

I wasn't in school that Friday because the semester had ended, and I had added more days to helping the golf business as the season was ramping up for the summer and I needed tuition money for the upcoming fall. I arrived bright and early for work at my friend's house, where his business was based, and before we could begin to move golf balls around to prepare them for delivery, we had to bring the 970 pounds of recycling out to the curb. Usually the recycling guy came around 11 AM or so, and we had finished putting the material next to the neighbor's one bin by 8 AM. We then set to work doing inventory in preparation for the end of the month. There would be no driving to golf courses that day. We broke for lunch around 1 PM and noticed the recycling was still stacked out front. After a quick sandwich, we were back to work and when we started to wrap things up around 5 PM the recycling sat exactly where it had been since 8 AM. In the kitchen, cracking a beer open to officially kick off the holiday weekend, we heard the unmistakable sound of the recycling truck as it turned into the side street where we were. We expected the truck to stop, but instead we heard the truck accelerate and we watched as the truck whizzed by the house. We swung our heads towards each other as we realized that the recycling guy was blowing us off, unwilling to stop and pick up our almost half a ton of material. It was hard to blame him. It was hot, it was after 5 PM on a holiday weekend Friday, and he just wasn't getting paid enough for this shit. No one would ever know why the material didn't get picked up. Except we knew. Throwing our beers down on the counter, we raced outside and into the street gesturing wildly for the recycling guy to stop. Almost a block away we caught him. He tried to pretend he didn't know what we wanted, but he knew we knew and the negotiation began. A small amount of begging, the offer of assistance, and a frosty six-pack convinced him to back up and return to the house. Fifteen minutes and a hundred apologies later, the recyclables were finally on their way to a new life and my experiment was over. To the recycling guy, thanks for being a good sport. That was way above the call of duty. Recycling can make you crazy like that.

Chapter 5

My new job required me to move from the Seacoast area north of Boston to southwestern New Hampshire. I was married at the time and my spouse had secured a teaching job seventy miles north of Keene. The plan was to move to an area equidistant for the opposite direction commutes, and purchasing a house was on the horizon. I had two weeks after graduation to relocate before I started my job. Enlisting a very tolerant friend, the same indulgent friend who helped me move 970 pounds of recyclables to his home for curbside collection on Friday of Memorial Day weekend, we rented a U-Haul and moved our meager possessions into my in-laws' basement—where, in the dark of night in June 1992, my friend and I were confronted by the local police, who had been alerted to suspicious activity. Weapons drawn, flashlights blinding us, they forced us to drop the couch we were moving *in* to my in-law's basement. It took a little fast talking to assure them we were not, in fact, burglars and we finished the chore and set off for our new geographical home—a summer sublet on a mountaintop thirty minutes north of Keene that had its own micro climate. It lay at the end of a mile-long gravel road that was heavily rutted and featured a gauntlet of not-too-friendly pit bulls, making any pass down the road in something other than a car a virtually impossible trek. The nearest store was eight miles away, and it didn't feel as though we'd exactly "arrived" to start this new adventure. The evening before my first day, a lovely Sunday summer eve, my then spouse and I drove what would become my daily commute into Keene so she could get a look at my workplace. I had been to the dump once before, but my head, at the time, wasn't absorbing the environment as much as I was thinking about some of the things I wanted to achieve, besides making sure no one else got murdered.

The dump was closed on Sunday, and it's one of those places you'd never go to unless you had to go. Pretty much every dump that ever was is located right on the border of another town. I liken it to a town-versus-town universal raspberry, because the location of the dump was never because of desirable geographic features. In fact, up until the 1970s, the Keene dump had a stream that literally ran through the middle of it before it was re-routed to improve the dump more than it was to improve the stream. The rumor was that the owners of the property, which was leased to the city, thought they could improve the value of their property by filling a ravine which, when the optimum height was achieved, would produce an unencumbered view of Mount Monadnock, which recently surpassed Mount Fuji as the most climbed mountain in the world. Meanwhile, there was a big hole to fill, and Monday through Saturday much of the waste generated in the southwestern portion of New Hampshire was deposited there. I was taking over a dying breed—the unlined landfill. The Environmental Protection Agency, through the Resource Conservation Recovery Act, was working with all the states to shut down all unlined landfills because of the threat they represented to contaminating underlying groundwater. Many communities, like Keene, were talking about saving the planet with the most recent wakeup call being the 112-day odyssey of the *Mobro*, also known as the Garbage Barge.

Leaving its berth in March 1987, the *Mobro* was loaded with over three thousand tons of trash from Islip, New York. It's original destination was Morehead City, North Carolina, where an enterprising entrepreneur connected with a Long Island mob boss and made a deal with a project that sought to generate methane gas with the decaying refuse. Arriving in North Carolina in early April, the *Mobro* was eventually turned away after the local press was alerted to the scheme, and it began to cruise the eastern seaboard for a home. With no US ports willing to allow the *Mobro* to dock, the barge sailed towards Mexico, where it was prohibited from entering Mexican waters. Belize was next on the list of rejected ports, until the *Mobro* finally turned around and sailed back to New York. After a contentious legal battle, the three thousand tons of trash was finally incinerated in Brooklyn, and the ash ended up back where the trash originated, buried in a landfill in Islip, New York.

A manufactured crisis of sorts unfolded between 1982 and the *Mobro* barge incident, when over three thousand municipal landfills were closed and the "not in my backyard" syndrome, followed by BANANA (build absolutely nothing anywhere near anyone), prevented new municipal landfills from opening. Reduce, reuse, and recycle became the battle cry of environmental activists, as significant fractions of the waste stream were thought to be able to be diverted from incinerators or landfills one bottle, can, and newspaper at a time.

Large, corporate-owned landfills replaced the dumps and the cost of disposal began to soar. What had once been free now had a cost associated with it, and if there is a universal truth, it's that people do not like to pay for things they no longer want. They just want it to go away. In 1992, when I began to oversee the Keene landfill, there was room for five more years of stuff, maybe seven years if an aggressive diversion program was instituted.

Driving down the access road on that quiet Sunday, I noticed a lot of birds just on the horizon. And then I saw the sides of the road. Maybe you remember the Keep America Beautiful commercials from the early 1970s. Ironically the crying Indian commercials, as they came to be known, which were played by an Italian character actor named Espera Oscar de Corti, began to make Americans aware of their penchant for littering. But Espera would have shed a lot more tears if he had seen what I saw then. Sure it was litter, but on a scope and scale that is almost impossible to imagine. All along the quarter-mile access road were layers of windblown plastic and paper from the landfill extending yards into the underbrush.

All through the branches of the immediate forest were plastic bags tangled, mangled, and shredded. The landfill, or dump, still loomed ahead. The birds came into focus. Seagulls. How could this be? We were one hundred miles from the ocean. There were upwards of a thousand seagulls circling in the sky or busy foraging through the thin cover of sand that barely covered the recently buried refuse. Surveying the scene, I took stock of what I had just inherited. A twenty-acre landfill, a two-car garage that housed the recycling center, a decrepit office trailer, an attendant shack, and of course, a porta-potty. Ringing the landfill and recycling center and everywhere I could see was litter that made the access road look pristine. It was an assault of the senses. Aesthetically awful, olfactorily offensive, and even what should have been a quiet Sunday was pierced by the cries of countless seagulls. I turned slowly in circles drinking it all in. A lot of things crossed my mind, and I stayed silent. To create order of this chaos was a tall directive. This was going to be a marathon, not a sprint. I knew I would have my fill of this place and I just wanted to get out of there, to forget about how overwhelming it all was for the next twelve hours before my alarm would go off summoning me to this place that would be my home away from home. Closing the door of the car, shutting out the sounds and the smell, I turned the car around and began the sobering drive back out the access road. About one hundred yards before the intersection of the road my then spouse turned to me aghast, "My God," she shrieked. "This is a dump! You work at a dump!" There was no disagreeing. It was bad. And as I turned on to the main road, leaving the shithole behind, I knew one thing for certain. That starting the next day, it wouldn't get any worse. Hello, my name is Duncan Watson. I'm with the government. I'm here to help.

Chapter 6

My first day at the dump, June 8, 1992, gave me a rude glimpse of the challenges I faced. The existing recycling center was not much more than a dilapidated two-car garage about thirty feet wide by forty feet deep with twelve-foot ceilings. The garage doors could not be closed because the recyclables were spilling out the doors. Standing a mere two feet inside the doors were six workers standing waist-deep in the recycling. Scattered across the front of the building, where the garage doors would have closed if they were able, were eight Gaylord boxes—an industry term coined by the Gaylord Container Company of St. Louis where these bulk boxes were first manufactured. A Gaylord box is a cube-shaped, three feet on each side, triple-wall corrugated box usually constructed of heavy-duty cardboard. Twenty-seven cubic feet makes a cubic yard. They are big enough to hold a significant amount of product and when empty can be easily manipulated by one person. When full, they fit nicely on a pallet, and if loaded correctly can be stacked up to three high. Usually, a Gaylord box is used as a temporary storage container. Depending on the material and the density when compacted, the Gaylord containers would be filled and set aside until enough Gaylord's were available to produce a bale.

Those first hours, I simply watched to see the work crew in action. Probably because it was my first day on the job, the crew started on time. Wading into the recyclables that literally filled the building, each person found a spot because they could make no further forward progress due to the wall of recycling that blocked their way. When I say that the building was full to the rafters with recycling I'm not exaggerating. The building was so full that the side walls were bulging with the volume. A morbidly obese building. Going back to the volume contained in a Gaylord box, a quick bit of math revealed that there was about 14,400 Gaylord boxes worth of material inside the building. At the rate of production that first day, there was no way it could ever be processed. I watched in fascination as the workers bent down, picked up one piece of recycling—a can, a plastic milk jug, a glass bottle—stood up and scanned the Gaylord boxes for the corresponding material, tossed it in, or at least they made an attempt to hit the target as the toss could be upwards of twenty-five feet. I estimated their accuracy at about 80 percent. Containers were flying in every direction, sometimes resulting in midair collisions with the occasional explosion of two glass bottles shattering on contact. Whenever that happened, a sarcastic cheer would erupt from the crew, as this was a satisfying respite from the monotonous job. It seemed to take an impossibly long time to fill up a Gaylord given the ready feedstock, but when a box was finally filled, all work would cease while one worker would get a fork truck to remove and store the container for later baling. The crew would continue to wait until the same guy who removed the Gaylord for storage came back with an empty Gaylord. Why the production ceased or why another worker didn't set up a new box was a bit of a mystery, but it reminded me of an old stereotypical joke about public works employees: Two public works employees are sitting in a truck when the driver slams on the brakes, jumps out of the truck, grabs a shovel from the back and proceeds to beat a turtle to death. When the driver gets back in, the passenger asks, "What did you do that for?" "Because he's been following us for the past hour," the driver answers.

And so I had my own real-life version of grotesque inefficiency. From time to time, a recycling truck that had been out picking up curbside recycling in town would arrive and deposit new recyclables that obliterated any progress that the workers might have made over the past hours. In fact, they were making no progress at all. It was apparent that the influx of new material outpaced their ability to process what was already there. Of course, the evidence of the inability to process was obvious by the fact that the building was full to bursting. But if the battle was being lost at this pace something didn't add up. It was at the very end of the day, the recycling spilling out six feet beyond the doors, when I got my answer. The foreman radioed to Bob, who came off the landfill with the bucket loader and its six-cubic-yard bucket, lined up the loader, dropped the bucket, drove into the pile of recyclables scooping up a full bucket, backed out, then took a slow ride up to the landfill so that nothing spilled, then unceremoniously dumped the recycling on the ground like so much refuse. Aghast and appalled, I watched this happen three more times until the garage doors could be semi-closed, signifying closing time.

If there is one thing I cannot abide in this profession, it's breaking what I see as a contract with my customers. People from all over town had gone to the trouble to recycle— dutifully preparing the containers, rinsing them out, the overly ambitious unnecessarily removing the labels, and putting them out on the designated collection day. They expected, reasonably, that their efforts were not for naught. If they saw what I was witnessing, it might discourage a lot of people from recycling . . . permanently. Just before he left for the day I told the foreman that was the last time he was to radio Bob to cart recyclables to the landfill.

Tuesday morning, my second day at the dump, was the first day that recycling would not be buried in the landfill. I had bodies to work with, but not much else. One of the nice things about working at a dump is that it's almost an ideal place to channel your inner *MacGyver* (a popular TV show from the mid-1980s that featured a guy who could fashion something useful from seemingly nothing). Within an hour, the foreman and I had all the materials I needed to construct a sorting line. Most sorting lines are mechanized, but this one was stationary and non-kinetic. I found eight fifty-five-gallon drums and lined them up end to end. I laid three-quarter-inch plywood on top of the drums, and then nailed two-by-fours around the perimeter of the plywood, leaving an opening at the end. It was effectively a three-sided tray. I butted the open side of the tray against the side of building, and cut a small opening in the siding of the building. Outside the building near the hole that had been cut, I made a horseshoe-shaped wall using concrete blocks, which are often called mafia blocks due to the penchant of the mafia to embed an enemy's feet in concrete to better weigh them down before disposing of them in a water body. I put Gaylord boxes two deep in front of the plywood tray separated by fifty-five-gallon drums that were fitted with glass crushers mounted on the rim of the drum. Four people were assigned to work the sorting line, and two people were assigned to feed the sorting line using thirty-three-gallon plastic trash barrels. The two feeders would shovel recyclables into the thirty-three-gallon bins, then carry it over to the sorting line where it would be dumped out onto the line and the sorting process would begin. Each Gaylord was assigned for a specific material and the recyclables got pushed down the line until the sorter who had that material picked it off the line and tossed it into the corresponding Gaylord. After a few barrels had been sorted, one of the workers would take a push broom and push the small fragments of broken glass and non-recyclables through the opening in the building. When the residual pile built up enough, Bob would be called down to bucket it away. Now only trash was being buried in the landfill, or at least we were not treating the recycling we received as trash. The difference in productivity was astounding. Not only were we able to keep up with the daily deliveries, but we started to peck away at the mountain of recycling. Eventually, we could finally see the back wall of the building after six months of toil. The building remained bowed even after relieving the pressure pushing against it, a reminder of where we had been. The sorting scheme seemed to motivate the crew, and the days became measured in production. Small improvements were added as the days blended into weeks and months. As the weather turned colder, production actually increased because the aerobic activity of flinging recyclables into Gaylord boxes helped keep them warm in a building with zero heat.

This primitive sorting system was a promise of things to come. There was an intention to build a materials-recovery facility, but the plans were about 20 percent complete when I arrived and the construction was not yet funded. The current two-car garage was going to be home for well over a year. Turns out that an impending landfill closure is a good political motivator, and being in the "Live Free or Die" state of New Hampshire meant that the hometown politicians wanted as much local control as possible.

The simple fact was something had to be done because the convenient landfill was soon to be full and the days of just finding some crappy land to use as a dump were long gone. Recycling was also getting a lot of attention as a tangible way that every citizen could lend a helping hand to save the planet. Communities across the United States were upping the recycling game, driven largely by the rapid growth of China's economy and their insatiable demand for raw materials. Corporations started investing in infrastructure to produce more goods from recycled material, and Keene began to coalesce around a plan to become a tiny spoke in a giant wheel.

Funds were committed to construct a new materials-recovery facility (aka, recycling center), and the engineering and design for the new facility got put on a fast track. It was going to be fully eighteen months before a new facility would come on-line, which prompted a lively political discussion on who would be best suited to run this new materials-recovery facility. This was during the height of a privatization movement that was sweeping through municipal governments. Politicians with an ideology that the private sector could do everything better than the public sector got elected and started to target what they thought were operations fraught with waste and inefficiency. To appease the privatization zealots, it was decided to put the operation of the soon-to-be-opening materials-recovery facility out to bid. A request for proposals was drawn up with a proposed scope of service, and the private sector was now looking at a wide-open door to take over a publicly funded, three-million-dollar facility. It was my feeling that the city's Solid Waste Division, of which I was in charge, should also have an opportunity to bid on the operation. Everything about the bids presented would be transparent, and frankly I thought we could be every bit as efficient as the private sector, maybe even more so. Most municipalities fund their solid waste operations via the general tax rate, but in Keene, the operation is funded through a special revenue fund whereby the charges for disposing trash and the revenue from the sale of recyclable commodities fund the entirety of our activities. We ran our operation like a business, and while we didn't pay any taxes on "profits," it was clearly stated in the request for proposal documents that because this was a publicly funded facility on city property, that neither the public nor private sector would pay any taxes on anything. Even though my work force was generally paid a bit less than what the private sector paid, the cost of providing benefits, primarily health insurance, made my cost of labor considerably higher. Still, I believed that our efficiencies could tip the balance in our favor, plus we had the skills and knowledge because we had been running the recycling center and it seemed reasonable that we should have an opportunity to prove our worth.

I took the lead in responding to the bid and effectively guaranteed that we would perform the work proposed. One private-sector proposal was received and the two proposals, the city's and the private-sector proposal were sent to a third-party committee for review and recommendation. The city's proposal was strong. We had had several years of experience running the city's recycling program and despite the fact that the new materials-recovery facility had not yet been completed, my estimates for costs were based on operational experience and reasonable estimates of things that anyone operating the facility would have to pay, such as utilities. The third-party review team recommended to the city council that the operation of the material recovery facility be awarded to the city Solid

Waste Division because, among other things, our proposal was $97,000 less expensive per year than the private-sector proposal.

I'll never forget the city council meeting where the decision on who should operate the materials recovery facility was made. The debate lasted hours, with the ideologues arguing passionately that it just was not possible that the public sector could do it cheaper than the private sector. I was accused of being a liar and of hiding facts and figures in spite of everything in my bid being available for scrutiny. The mayor finally called for a rollcall vote and I practically held my breath as each councilor, with their own soliloquy, announced their vote. The final tally was 8–7 in favor of awarding the bid to the private sector. I was stunned. A lot of different thoughts raced through my head. Did my crew and I just lose our jobs? Why would they have awarded the bid to the private sector when the city's bid was considerably lower? How am I going to pay for my student loans? How am I going to pay my mortgage for the house I recently purchased? This just seemed overall to be a gross miscarriage of political decision-making, but I was really powerless to do anything about it.

I was not alone in my characterization that the vote was a miscarriage of political decision-making because it turns out the city operation not only had fans, but fans who were passionate about keeping the public trust, and accountability. As a core group of these fans mobilized to determine what steps they might be able to take to right what they perceived as a wrong, they learned that there was a parliamentary procedure whereby any one of the city councilors who voted in the majority decision could, within two weeks of the original meeting, make a motion to reconsider the original vote. And that's where they turned their attention—towards the eight city councilors who voted to award the bid to the private sector. It was a political shitshow for the two weeks until the possibility that the vote could be reconsidered. The local radio talk shows were alive with conversation about who was best to run the recycling center, several of the activists set up a table in front of city hall to gather petition signatures to encourage the city councilors who voted in favor of privatizing to change their vote. It was all akin to *Mr. Smith Goes to Washington*, the classic little guy versus big guy movie starring Jimmy Stewart as a naïve, but well-intentioned freshman Congressman going against big machine politics. As for me and my crew, there was a lot hanging in the balance, and there was no way to know if there would even be a vote to reconsider, let alone to guess how things might turn out.

The city council chambers was packed to the gills, with about 95 percent of the gallery in favor of awarding the bid to what was characterized as the lowest-cost responsive bidder, which is generally how government bids are awarded. One of the city councilors did put forward a motion to reconsider the vote and again a vote was held where each city councilor stood up and issued forth a several-minute monologue articulating why and how they intended to vote, then putting forth their vote on awarding the bid. The final tally was 8–7 to award the contract to the city's Solid Waste Division. There was indeed jubilation that day. I believed virtue had been served and now I had a chance to really show the community what we could do.

The stress of the political battles I was fronting was exacerbated by a tobogganing accident that laid me up for twelve weeks with a pinched sciatic nerve. If you've never experienced a pinched sciatic nerve consider yourself lucky. Imagine someone plunging a serrated knife into your ass, and every time you breathe the knife is twisted just to remind you that it's there, then you have a vague sense of its agony. I was bedridden for eight weeks and had to use a walker to even be remotely ambulatory for four weeks, and I did my best to conduct business with my laptop and a telephone, staying as still as possible to avoid the twisting knife.

In my absence, a pole barn had been erected adjacent to the two-car garage that had been our home but that was now being disassembled before being reassembled farther up the hill where it would serve as the state's first dedicated household hazardous-waste collection center, as well as a good example of reuse. Several of the straightest trees from the recently cleared site of what would soon be the new material-recovery facility were stripped of branches and buried six feet into the ground. Trusses were placed on top to tie the poles together and a corrugated metal roof was installed. Lastly, the structure was sheathed on three sides to, in theory, withstand the wind, rain, sleet, and snow that was to come.

It's difficult to really understand how challenging it was to process recyclables in New Hampshire in the winter with only ten millimeters of plastic sheeting separating the workers from the elements. My office trailer had been moved across the driveway that went up into the landfill, and my view was of the bone-chilled men who were constantly moving inside their plastic igloo to keep from getting frostbitten. While my trailer had heat, it never got much above 50 degrees. I was very thankful to be back at work after my twelve-week sciatic nightmare. I had been going stir crazy cooped up in bed at home, and there were so many moving pieces happening simultaneously that I really wanted to be on the ground managing things. I was still quite fragile though, and I was using a kneeling chair at my desk to take the pressure off my back. That first day back, the building that had been the recycling center and would soon be the household hazardous-waste collection center was slated to be moved up the hill. The steel frame intact, what remained of the building (all the siding had been removed) was jacked up and a tractor-trailer backed under the bracing before being lowered down on the trailer to begin the climb uphill.

I was focused on some computer reports that I was catching up on, when, glancing up, I noticed the building was gone. I got up from my desk gingerly, my back constantly reminding me that it still needed attention, walked to the door, opened the door, and stepped onto the landing of the staircase. Except the landing wasn't there. I stepped into space and quickly got a lesson on gravity. Gravity, the natural phenomenon by which all things with mass are brought towards one another. In this case, because of the physical action of stepping forward with my right foot onto the landing that was not there I suddenly pitched forward on a trajectory that had my face accelerating toward the ground at an alarming rate of speed. I was, in fact, shy of the twenty-two miles per hour speed because my descent was less than a second, but even factoring in wind resistance, the fraction of a second that it took to complete my four-foot fall ended in a sickening thud. I was heaped on the ground howling in pain while simultaneously going through a full system check to determine just how bad it was. The first thing to strike the ground was my nose followed by the rest of my 190 pounds. I felt the blood coming from a gash across the bridge of my nose, and my back was very unhappy with me, but more than anything I was terribly embarrassed. As I lifted my head I noticed Bob, the landfill compactor operator, standing across the driveway, coffee in hand, with his mouth wide open. He had witnessed the whole thing and now he was caught in the dilemma of feigning horror while trying not to guffaw because it must have been a hilarious sight—the door to the trailer comes banging open followed by his boss stepping into thin air, falling face first into the dirt and gravel below. It was slapstick of the highest order.

With my system check complete I rose slowly as I came to a full boil. My first question: Where the hell are the stairs? Unbeknownst to me, the guys moving the old recycling building had to make a wide arc to turn the corner in order to move the building up the hill. That wide arc necessitated removing my staircase. It isn't difficult to imagine the comedy of errors of not informing the person occupying the trailer that there was no longer a staircase, but of course I found out quickly enough. I suppose it could have been worse. I could have been one of the workers frozen to the core flinging bottles and cans in a vain attempt to keep warm, the wind pounding the plastic sheeting to thunderous effect for eight hours a day. I dusted myself off sputtering about why I wasn't told about the stairs, located a step ladder, climbed back into the trailer and got back to work.

Chapter 7

Crazy abounds in the world of recycling. There are many, myself included, who take recycling to an extreme. Of course, being in charge of a recycling program for a community of twenty-three thousand is bound to produce more than a few eccentric customers. I received a call one morning from a breathless woman who was so excited by her question that she simply couldn't articulate what it was that she wanted, so I suggested that, if she had time, I would be happy to meet her in person to figure out what was bothering her. A few hours later, she showed up. Now I realize that one should not judge a book by its cover, but honestly this woman looked a little touched. If there is one thing I learned from years working in the hospitality industry, it's that customer service is a premium. With all the professionalism I could muster, I reached out my hand to greet this tousled-haired woman with eyes so wide open that I'm certain young children gave her a generous berth. She launched into a conspiracy theory rant right away, and I tried to follow the increasingly thin threads that she was trying to connect. She implicated the NSA and the CIA for monitoring her and interfering with her brain waves, and to be truthful there was 2 percent of me that fully believed everything she was saying and that it all made perfect sense when she finally arrived at the actual reason for her visit. She wanted to know if she could recycle used SOS pads and wire coat hangers. She needed the answer to be yes, and while it's technically true that most everything can be recycled, there are a few asterisks that lead to less desirable answers, such as having a critical mass of homogenous material, storage space, and ultimately a market to ship the material. And, because recycling is a business, someone has to make some money every step of the way. In this case, the wire hangers were pretty straightforward. They are made of metal and can reasonably be blended in the scrap metal pile. If a magnet sticks to it, it's ferrous metal and is infinitely recyclable. The used SOS pad was a little more complicated, as I thought SOS pads are essentially steel wool, but I didn't think of it as ferrous metal. Now, in truth, I gave her the answer she needed. I told her to put her wire hanger in the scrap metal pile and the used SOS pad in the recycling bin and that we'd take care of it at the recycling center.

We have what is known in the industry as a MRF, pronounced "murph," which stands for materials recovery facility. Back in 1994, the city of Keene opened what was and still is the largest municipally operated MRF in the state of New Hampshire. Now, if this statement causes you to shrug indifferently then you are not alone. Just because it's true does not make it particularly interesting. But, for those of us in this business, what I now had to operate was a Ritz Carlton in a sea of Motel 6's. Believe me when I tell you, we were the envy of thirty or so people who knew enough to care that that made me the king of the best kingdom. Having the largest municipally operated MRF in the state of New Hampshire, current population 1.3 million within its 9,350 square-mile border (for comparison purposes the city of San Diego has 1.3 million people within its 372 square mile border), simply meant that we had a factory that took in mixed resources. The resources that we accept, within a fairly narrow scope of specifications, are processed through the means of machine and human sorting, producing the raw materials (commodities) which are sold to be made into new products. For example, a glass jar can be endlessly recycled into a new glass jar, a steel can endlessly recycled into a new steel can, same with aluminum. Plastic, on the other hand is a lot more complex.

The plastic industry has done a great job (scandalous really) marketing their material as recyclable. Look at almost any plastic product and you are likely to find a number inside the recycling symbol, which would lead any reasonable person to surmise that the package or container is therefore recyclable. Technically this is true, but I go back to what I said earlier. I need a critical mass of homogenous material, storage space, and a market. I can easily accomplish those three criteria with the ubiquitous #1 and #2 plastic, but for #3, #4, #5, #6, and #7 plastic, that is often not the case, and in my small community, gathering the critical mass of plastic for anything other than #1 or #2 is currently impractical. It takes a couple of days to gather the 40 cubic yards of material to make a bale of #1 or #2 plastic. Earlier I referred to the ubiquitous three feet wide, three feet high, and three feet deep box known as a Gaylord container. I need forty Gaylord's to make a bale of plastic. In order to efficiently ship a product I need 24 bales of material, equivalent to 960 Gaylord's of homogeneous plastic. It would take years to accumulate enough material to ship any plastic other than #1 or #2. As an aside, I am still asked, with surprising frequency, whether the thing in their hand, something other than a typical recyclable, is recyclable. When I give the answer, which is usually "No," the next question is "Are you sure?" This is a weekly conversation IN MY OWN HOUSEHOLD. Yes, I am, in fact, sure. I feel compelled to note, as I will discuss at the end of this book, that in the case of recycling "stuff" that is currently not recyclable is primarily a function of volume. When economies of scale are achieved, such as the next generation of materials recovery facilities processing fifteen hundred tons per day, a lot more items become economically recyclable.

Paper products, which I often refer to as fiber, are also easily recyclable and represent a significant fraction of the waste stream. Paper products, including cardboard, or in recycling lingo OCC which stands for old corrugated cardboard, is made into new paper products. The reason I refer to it as fiber is that all paper products consist of paper fibers. The longer the fiber, the higher quality of paper can be made from that fiber because the length of the fiber makes a stronger, more durable product as the longer fibers weave together. Try to remember this as you shred every single piece of mail that comes to your house or every piece of paper at your office in an attempt to protect identity or confidentiality. Sure, shred those documents that have account information or social security numbers and the like, but by overzealously shredding all your paper into confetti you are effectively ensuring the only recyclable product that can be created by the dramatically shortened fibers is an egg carton. That's pretty much the end of the line for paper recycling. Beyond an egg carton, the fibers become so small they pass through the screens designed to sort the longer fibers from the short ones; they have no practical use and are now destined to be buried or burned as the solids are removed from the water in the wastewater treatment process.

MRFs are designed as quality-control stations in the chain of recycling. At the Keene MRF, material collected at curbside or from individuals dropping stuff off is separated into three main components: cardboard, mixed paper (if it tears, put it in mixed paper), and containers. Containers consist of glass food and beverage containers, steel cans, aluminum cans, and #1 PETE and #2 HDPE plastic bottles. The containers are mixed in a large hopper and mechanically conveyed up a steep incline which meters the containers by having the material not caught by the cleats tumble back into the hopper. This is so we don't end up with a huge slug of containers dropped on to the sorting conveyor at one time because it makes it much more difficult for a human to handle that much material. Next comes a shaker screen. This is not much more than a large vibrating sifter that screens out material smaller than two inches, which is to say, predominately broken glass, and the shaker helps to further meter the material on to the sorting conveyor. Now the humans get involved in the act. Three or four people, depending on the speed of the conveyor, look for the material they are responsible for. Station one gets colored and clear #2 plastic and clear glass, station two gets any #2 plastic and clear glass missed by station one as well as brown and green glass, #1 plastic, or

PETE (polyethylene terephthalate), which is most commonly associated with soda bottles. Whoever is selling #1 PETE plastic to manufacturers of food products is doing a helluva job because today more and more products are packaged in PETE. Station three pulls off the redemption containers, also known as nickels because they can be turned in for the deposit. We'll get to redemption containers a bit later, as New Hampshire is not a bottle bill state, but the nickel business is alive and well in New Hampshire. Station three also collects anything that got missed at station one and two to make sure everything that can be recycled gets recycled. At this point the machines take over. The only things that should be left on the line are steel cans (ferrous), aluminum cans (nonferrous), and items we don't accept (trash).

We hope that a combination of public education and the general intelligence of the recycling public captures the attention of the wishful and reluctant recyclers of our community to maximize diversion and results in less than 10 percent of the material we process going off the end of the line to end up as trash. We call the end of the line "residuals," and it consists primarily of broken glass, plastic bottle caps, and the more than occasional bowling ball or prom photos. Any odd item we've ever received at the recycling center is due to the wishful recycler.

I have personally witnessed attempts at recycling a live snake, a lawn mower engine, and an onion, which one of my considerate colleagues chopped up and served with my hot dog, unbeknownst to me, during one of our reasonably frequent lunchtime cookouts. Those early days had me working the line every day and it was exactly like that classic "I Love Lucy" scene where she worked at the chocolate factory with a very fast conveyor belt. As my arms windmilled, throwing things off the line into the correct bin, my mantra was "I have a master's degree, I have a master's degree." Humor is a great coping mechanism.

After the removing the nickel containers, the material travels under a cross-belt magnet, and anything magnetic gets carried off the line. This is where the used SOS pad comes back into play. If it really is ferrous, it will get carried off the line to be recycled with the steel cans. If it's not ferrous, then it won't get recycled and will drop off the end of the line into the trash. Ninety percent of what comes into the MRF gets recycled, and I can live with that. Now all that remains is aluminum and trash. Just before the aluminum cans reach the end of the line, they run across an eddy current separator. An eddy current separator uses a magnetic rotor with alternating polarity spinning rapidly inside a non-metallic drum driven by a conveyor belt. As nonferrous metals pass over the drum, the alternating magnetic field creates eddy currents in the nonferrous metal particles repelling the material away from the conveyor. While other materials drop off at the end of the conveyor, the non-ferrous metals are propelled forward literally leaping off the line and over the trash container into a bin filled with aluminum cans. Thirty years of watching this machine and I still find it fascinating to watch how it works.

The reason we do all of this quality control is because of the SOS pad woman and people like me. The wishful (overzealous) recycler is the pest of recycling professionals everywhere. My name is Duncan Watson, and before I knew better I was a wishful recycler. It started when I moved from the town where I conducted my recycling experiment to a town on the north shore of Boston that had a private recycling facility where one could drop off recycling. Once a month or so, I'd load up my car and drive my stuff to the auto salvage yard that had added a recycling drop-off center to try to take advantage of the rapidly expanding global commodities market, because there was real money in these discards. One day I arrived to find the facility closed. I think I just got there too late, but all the bins were accessible, so I unloaded my stuff anyway. They had separate bins for everything and I was more than happy to comply with their straightforward system. Except I took the system too literally. You see, at the bottom of my recycling bin I had a lightbulb. A clear lightbulb. I had put the lightbulb in there. I reasoned, logically, that because this was a clear lightbulb, the lightbulb was made of glass, of course it belonged in the clear glass recycling bin. That's exactly where I tossed it, happily transported to my twelve-year-old self as it satisfyingly exploded into tiny pieces amongst all the clear glass bottles. Little did I know I had just ruined that entire load of glass. Probably forty thousand pounds worth. Yes, lightbulbs are made of glass, but the glass used in a lightbulb melts at a different temperature than glass food and beverage containers. They would have found this out during the bottle making process because the new food containers would have pieces of glass that never melted. The incompatible glass remains embedded into the container rendering it structurally deficient. If you are one of those overzealous people, like me, ask before you assume you know what you're doing. We'd rather the question than deal with the huge financial consequences of a rejected load. I'm still not sure if SOS pads are ferrous. One of these days I'll have occasion to find out.

Our brand new almost one-hundred-thousand-square-foot MRF opened in early 1994. Gone was the two-car garage. We had made it to Broadway. During one particularly contentious city council meeting, shortly before the MRF opened, one exasperated city councilor stood up and said disparagingly that I treated the MRF as my own personal kingdom, so, even if I might have thought that privately, apparently more than one person saw this. Of course that quote made it into the local paper the next day, for which I took endless grief.

In our municipal government the chief executive officer is the city manager, and I was proud to give him the full tour of our impressive achievement. The tour included an in-ground weight scale, conveyors, balers, big doors, and high ceilings. We spent over an hour exploring all the nooks and crannies. As we walked back to his car, he noticed a placard at the very top of the west side of the building no bigger than a street sign, thirty feet in the air. The city manager asked me what that was and I confidently explained that it was likely the manufacturer of the building putting their emblem on it. As he put on his glasses and focused his eyes he read slowly "King" . . . "Duncan's" . . . "Palace." I about fell over. It was a perfect gotcha. Needless to say, I have some co-workers with a sharp sense of humor. I don't think the city manager was amused, but that sign remains to this day.

Chapter 8

In July 1999, the process of closing the landfill began. Operating a landfill is extremely profitable and the three-million-dollar price tag to entomb more than 850,000 tons of waste below four feet of cover, including a plastic liner to prevent precipitation from coming in contact with the buried refuse, was already in the bank and then some. The trash pyramid was further shaped to a certain slope and grade—essentially for every two feet the landfill extended horizontally, it rose up one foot, also known as a 2-to-1 slope. The contractor finished the capping in early spring 2000 and the most important thing at that point was to establish a vegetative cover so the steep slopes didn't erode. Closing a landfill is a much-prescribed affair, and deviation from accepted specifications is rarely considered. In a small nod to artistic inspiration, the State of New Hampshire Department of Environmental Services approved a seeding plan to include hundreds of pounds of wildflower seeds. The manufactured topsoil that made up the final six inches of cover consisted of a mixture of sand, short-paper-fiber waste, and composted wastewater sludge also known by the newly coined word, biosolids. It was the perfect medium for growing most anything, but the wildflowers went crazy and within weeks the twenty-acre landfill was ablaze with a veritable kaleidoscope of colors. Arriving to work early one morning, I walked to the top of the landfill, the vista of Mount Monadnock looming in the distance and the valley below shrouded in a defined layer of fog. The sun was just rising, illuminating the mountain and the fog. Wildflowers were blooming everywhere, and in the soft breeze came the humming sounds of bees hard at work. It was hard to resist bursting into *The Sound of Music*, arms outstretched, twirling in a circle. The hill was alive and in all my worldly travels the top of this waste heap turned into the most unlikely member of the top three most beautiful places I'd ever seen.

Eventually, the combination of the closed landfill and the extremely cool, for the six-to-eleven-year-old crowd, mechanized recycling center became a staple of the elementary school tour circuit. It was a magical thing to watch as we concluded our tours with a walk to the top of the landfill where the children gathered bouquets of wildflowers to bring home to Mom and Dad. Job satisfaction during these moments was off the charts.

Another unconventional practice that was employed was using alternative mowing machines to keep the vegetative cover under control. One unwanted thing on a closed landfill is the possibility of trees or shrubs growing on the landfill and possibly breeching the plastic liner with their water-seeking roots. Four times a year was the suggested mowing schedule. I stayed late after work to watch the mowers unloading. The ramp of the tractor trailer was secured and the offloading of 199 sheep and one sheep dog began. Since we had what amounted to a twenty-acre pasture why not use it as a pasture? The original thought was cattle grazing, but when I contacted the County Agriculture Extension, they led me to a company that had sheep. If you've never seen a delivery of sheep, it is a sight to behold. Sheep are very vocal about what they think, expressed in bleats of various volumes. Before long, the shepherd, along with several border collies, efficiently corralled the sheep into a one-acre pen and in short order the sheep were drunk with happiness as they gorged themselves on the rich grasses that covered the landfill. The sheep were with us for several years, arriving in the spring and departing in early December. The sheep were entertaining, and they added a lot to the landscape as they worked their way across the twenty-acre pasture one acre at a time. By far the most interesting member of the flock was the sheep dog whose sole mission was to protect the flock from predators. The sheep dog assigned to this flock looked exactly like the sheep dog featured in the Looney Tunes cartoons. The only thing different is he didn't punch a clock to check in and out. The landfill was largely surrounded by woods to the west, and potential predators included coyotes, fox, bears, bobcats, rumored mountain lions, dogs, and humans. It was virtually impossible to distinguish the sheep dog from the flock as he blended in that well, but encroach within a certain distance of the flock and you would discover quickly that this dog meant business. I watched the sheep dog with his handler, the shepherd, and he was playful and friendly, but if he didn't know you it was readily apparent that he took his job seriously and he was not to be trifled with.

Of course, the sheep just added to the school tour frenzy, and we had to make sure the shepherd was there because everyone wanted to pet the sheep dog. He was very stoic about all of it. The only thing he was missing was a pair of shades because he was the height of cool. The sheep, well, they probably didn't appreciate the scenery because they constantly had their heads buried in their work. From what I could tell they certainly enjoyed their work.

Chapter 9

All told, the landfill, operating for over forty years, has close to a million tons buried within its twenty acres. Formal waste management efforts in Keene began in the early 1900s with the construction of an incinerator a half-mile from downtown. Known as the garbage destructor, the incinerator operated for a few years before straight-out landfilling became the preferred management option. It's actually harder to fully burn trash than one would think. The dump I operated originally opened in the 1950s but was closed for a while because of the effects of a stream running through the middle of the dump. The practical aspects of having a huge hole in the ground at the farthest edge of town produced a plan to reroute the stream, and in the early 1970s the dump was reopened and operated continually until it finally filled up in 1999. Some of the old timers would reminisce about coming to the dump with their fathers in the 1950s and driving way down into the hole to get rid of their trash. A combination of dumping solid and liquid waste, occasional burning of whatever would burn, and the added bonus of taking shots at the healthy population of rats that feasted on the discards almost evokes a days gone by Americana scene. Dumps have always been social gathering places and it was many a rite of passage to join your father, which has since morphed into more equal gender distribution, on a trip to the dump because it was always possible that something exciting was going to happen when you got there. If nothing else, countless people enjoyed the view.

One old-timer's name is Freddy, but when given the nickname Captain Cardboard, it stuck. Freddy, now in his late seventies and still working part time at the dump, was one of the workers I inherited when I took over the dump operation in 1992. At the time I started, I was the only salaried and benefited employee at the dump. With the construction of the new materials recovery facility slated for 1994, we started to transition some of the contract workers into full-time, hourly with benefits employees. Freddy was a good worker with a habit of actively wondering what his co-workers were doing when not under his watchful eye. Freddy also liked the occasional prank to spice up the daily drudgery. One day I was having a particularly intense meeting with the public works director in my office trailer when the door was slammed open and in walks Freddy, unaware that the public works director was sitting behind the door, and Freddy was wearing a woman's wig that he had happily fished out from the landfill. I gave Freddy the eye to let him know I was not alone in the trailer. He turned to face the public works director looking quite ridiculous and said a few words of small talk before beating a hasty exit.

The Captain Cardboard nickname became a thing because baling cardboard in the early days was a hand-loading affair that took hours to produce a bale and Freddy seemed to like that chore better than anyone. Now it takes a few minutes, as the loading is done with equipment, and you no longer need to hand tie-off the bale because it's done automatically by the machine. I don't know how much cardboard Freddy has processed over all these years, but it's surely enough to have earned the Captain Cardboard name, and I'm grateful for his friendship and the many miles we have traveled together.

Al was another one of the contract workers who was brought on as a full-time city employee. Al was born and raised in the area, and his particular skill was as an equipment operator. You've seen videos in your life of skilled equipment operators who can open a beer bottle with the teeth of an excavator or work a job where the hand-eye coordination and natural talent makes it seem as though the equipment is an extension of the human itself. Al had that kind of skill. When you have a multiple decade relationship with someone you get to bear witness to lives led. Marriages, divorces, a whole handful of children from the different marriages, Al had a very busy life outside of work. Through it all, though, he kept showing up to work and after thirty years, in 2023, retired from the dump. He elevated himself to a foreman, and up to his retirement day still had mad skills in any equipment he operated.

I was fortunate to know Al because he had all kinds of talents that I was not gifted growing up. One of those talents was as a builder. Now Al was no fine carpenter, but his father had taught him carpentry skills and I enlisted Al to help me build a screen room perched on ledges of the river I live on, as I had zero clue. I know the screen room was built in 1999 because my then two-year-old daughter would lament my dashing off after getting home to work on the room. She knew I was doing this work with Al, and she associated him with the sound of hammering, so my daughter named him Bang Bang Al, and that's how he'll be forever known in my family.

Donny is probably the most cheerful person you have ever met. And he is also among the most distractable humans I've ever met, and I was once married to one of those people where you'd be having a conversation and suddenly they exclaim, "Hey look, a squirrel!" Donny is a dedicated environmentalist and completely passionate about his work. He's bright, inquisitive, and unfailingly positive. He also bears an uncanny resemblance to Matt Damon. Donny has got to be approaching his forties, but if I was a clerk at a liquor store I would definitely card him because he just doesn't look older than eighteen, and it has been that way the entire time I've known him. If you've ever seen the movie *Up*, then you remember the plot where a grouchy old septuagenarian embarks on an adventure in honor of his late wife and along the way he meets up with an affable (with a capital A) dog named Dug. Dug's most memorable line is "I've just met you, but I loooooove you." Dug also happens to be highly distractable. Donny is a human Dug. And to associate Donny with Dug is meant as the ultimate compliment because dogs are awesome, Dug is awesome, and I appreciate Donny for exactly who he is, eccentric ideas and all. If you ever meet Dug, I mean Donny, you'll never forget his million-watt smile, and sunny disposition.

If you've ever heard or seen the comedian Jim Jefferies, his mind and his affect seemed to mirror Hank. Hank was only with us for a short time. We hired Hank as a contract employee to pound the trash for the final six months before the landfill closed. Hank was born and raised in Brooklyn, NY, and he was the literal stereotype of an old school Brooklynite. Italian by way of ancestry, Hank spoke with his hands and he was a particularly animated conversationalist. Hank had been a police officer in New York City for several decades before retiring to New Hampshire. He brought his libertarian, contrarian, pugnacious personality with him, and I adored him. If you said black to Hank he would say white. It didn't matter if he too thought that something was black, if you said it, he would say white just to get a rise. He was very smart in the common-sense department, but mostly he just loved to talk.

When not at work, Hank could often be heard on our local talk radio station calling in to argue with the host. And yet, while we didn't see eye to eye on much of anything, I have this overwhelming endearing memory of him as a man with an oversized personality and oversized heart. After his contract ended, we stayed in touch. It wasn't hard, as most days I'd hear him rattling on about something on the radio, but his heart was always on his sleeve and he would do these incredibly thoughtful things such as gifting my children savings bonds for their birthdays.

Hank's time on earth was tragically cut short in his mid-sixties. Heart issues finally got him, but he's alive and well in my mind and heart, and I'm so blessed to have orbited his planet for the time I had.

"A happy worker is a good worker." This was a phrase I heard thousands of times whenever I came across Bob Paddock. Bob had a decade-plus career in the Highway Division as a laborer before transferring to the Solid Waste Division. What a gift he was. Bob was just a straight-up worker bee. He absolutely loved being at work. He hated to go home. He would often ask if he could stay at work even if we didn't pay him. Yeah, "no Bob, when you work at the dump you get paid." Bob said to me numerous times that when he died, he wanted his ashes spread at the dump. He was completely serious about that. I wish I had a fleet of Bobs. That guy can work. Into his eighties, Bob worked with us. There came a point that his physicality started to become a liability, and Bob reluctantly agreed to call it a career. Bob passed in 2023 and I reached out to his family to see if they were aware of Bob's wishes, but I can promise that if they say yes, I will follow them knowing that Bob's spirit will be eternally happy. Meanwhile, we are creating a memorial to Bob at the recycling center that will likely involve a plaque and planting a tree. And a quick note to my children. I do not wish for my ashes to be spread at the dump. I'd rather be part of the soil base for a tree planted in the yard. I'll let you decide where to plant it, and what kind of tree, but as a hint I'm very partial to maple trees.

I've always felt that the sign of a full life is to have it filled with characters. In this regard, I'm a rich man indeed. It takes a unique character to work at the dump. It's quickly apparent who is and isn't cut out for this line of work. Those who have more than a little quirkiness tend to stick around, even until death do us part.

I first "met" John, the gate attendant, via phone. Early in my career, I spent a year as an assistant reservations manager for one of the large hotel chains. Phone etiquette was drilled into us from day one. So, when John picked up the phone on my first call to the facility answering with a profoundly gruff single word, "DUMP," I knew I had my work cut out for me. The gate attendant in those early days had a building the size of a telephone booth. There were windows on three sides that did not open and a window in the door that should have opened but was broken shut. There was a slanted workspace where the log sheets were kept and a stool because there was no room for a chair. I had once seen four people come out of the booth, which looked like something out of a clown car because whenever I was in the booth with John we were shoulder to shoulder necessitating a multipoint shuffle to turn around to exit what we referred to as "the shack." It was, however, very difficult to stay in the shack for very long because John was a chain smoker. For eight hours every single day he worked there, I don't recall a time he did not have a lit cigarette. Never mind that it was against city policy to smoke, never mind that the shack was mere feet from the active landfill. Do you know what is created by decomposing trash? Methane gas among other gases. A lot of it. Methane is colorless and odorless, but extremely flammable and I often imagined the ensuing fireball each time I watched him light up another cigarette. John was well into his seventies when I arrived, and he'd been working at the dump for several years already. One day when working with John on the log sheets, I dropped my pencil. We were still computerless in 1992, and everything was done by hand. I had to go out the door and hold it open so I could bend down to retrieve my pencil, but I couldn't see it anywhere. The light wasn't great anyway, the windows were yellowed from smoke, and there was no electricity in the shack, so I had to wave my hand along the ground to find the object I was looking for. It was then I noticed the floor was coated in a three-inch layer of cigarette ashes and my pencil had been swallowed up. All I could think is I was glad I was not John's lungs.

It took a long time to get John to change his greeting. I asked him very nicely to go from "DUMP" to "Good morning [or afternoon], Keene Recycling Center and Landfill, this is John." I explained that I was trying to change the culture and to get customers out of the habit of saying dump. For the most part, people now refer to us as the Keene Recycling Center and Transfer Station, but there are many that will always refer to it as the dump. I let go of taking it personally, and I use the term "dump" often as a homage to the traditional cultural reference.

The logbook that John kept consisted of hash marks for each and every cubic yard of trash or recycling estimated by eye. The sum of that estimation was how we billed our customers. We charged twelve dollars a cubic yard for trash, compacted or not; recycling was free, to try to incentivize the haulers to increase recycling. This was a multi-million-dollar business being conducted by making hash marks in a logbook. It was crude, but it worked.

While I might have been successful in changing John's phone greeting, I gave up enforcing the no-smoking policy, a decision I made after really getting to know John. Born and raised in the area, John joined the army after the outbreak of World War II. Eventually commissioned as a second lieutenant, John served with the 36th Infantry Division in the Italian campaign. He was awarded the Silver Star for conspicuous gallantry and intrepidity in action against the enemy, and in early 1944, John saw action in capturing Mount Maggiore, Mount Lungo, and the village of San Pietro. Enemy resistance was furious, and the winter weather equally formidable. It's difficult to imagine the brutal elements coupled with the terrifying sights, sounds, and smells of war. And this twenty-six-year-old, by this time first lieutenant, was part of a raging battle against the Bernhardt Line, which relied upon the 36th Infantry to secure a bridgehead across the Gari River, which had been erroneously identified as the Rapido River. The 36th attacked across the Gari on January 20, 1944, and were repulsed by the 15th Panzer Grenadier Division. After two regiments were fundamentally destroyed, the attack was stopped on January 22, the day John was raked across the chest with machine gun fire, earning him a Purple Heart. In those two days, the 36th suffered 1,681 casualties: 143 killed, 663 wounded, and 875 missing. John, in many ways, was one of the lucky ones. The commander of the US 5th Army, Lieutenant General Mark Clark, was criticized for having ordered such a difficult frontal attack and was largely blamed for the disaster. After the war, the US Congress, at the behest of veterans of the division, conducted an investigation into the causes and responsibility for the defeat on the Gari River. After learning all this, I might have purchased John a carton or two of his favorite smokes. It took a lot of prodding for this humble man to tell his stories. He was an American hero and I was honored to know him. He retired in his early eighties. It was the transition from the pencil logbook to the computer that finally forced him to retire. In the ensuing years, you have no idea how often I have wanted to answer the phone "DUMP." John died at the age of eighty-eight. He will forever be my hero.

When the recycling center opened in 1994, I had to hire some new staff to sort the recyclables. The economy was starting to come out of the recession of 1992 and municipal employment was seen as an attractive option given the basic requirement of an entry-level job was a high school degree. The day I first met Kirk, I was convinced he would not be able to withstand the rigors of the job. Kirk seemed more suited to a poetry career than being in the middle of the cacophony that is the hallmark of a mechanized recycling center. I took a chance on Kirk for completely self-serving reasons. He would be my lunchtime savior. I expected him to be a weak link in the physical aspects of the job, as he was of slight build and he had the look and comport of an intellectual. But he ultimately proved me right and wrong. He had graduated high school, but he was one of the most learned men I've ever known. He was insatiably curious and an unparalleled raconteur. He was interested in recycling and resource conservation, and he had informed opinions on practically any topic that arose every single weekday during lunch. Political discourse would morph into the minutia of popular culture. Whether it was books or movies or the headlines on National Public Radio, he was always one step ahead of me. His humor was witty and clever. We shared an affinity for Monty Python and we would often bust out a favorite Monty Python quote that only we appreciated, belly laughs ensuing.

And for five years Kirk proved me wrong on his ability to hold up the physicality of the job. With his profound intelligence, he was efficiency in motion, and I was privileged to work side by side with him. Kirk took ill one fall and was hospitalized. I visited him in the hospital and while I was in my mid-thirties, Kirk seemed to belong to the same immortality club as he was a mere forty-three years old. And just like that he was gone. I later found out he died of complications of hepatitis C. His family claimed Kirk mentioned he had been pricked by a needle while sorting recyclables. It's certainly not impossible, as needles are a common sight to those who work on a sorting line. He never mentioned the needle stick to me. His family received a small insurance settlement, which hardly makes up for the loss experienced by his wife and son. I lost one of the better workers I've ever had, and more significantly I lost a friend.

On a completely different end of the spectrum was Bob Farnsworth, the operator of the landfill compactor. Bob had been pounding trash for years before I arrived. The machine he operated is formidable with 560 horsepower, 7-foot-high by 5-foot-wide wheels with 40 hardened steel cleats welded to each one. The cleats provided traction as well as compaction. The compactor weighed 61 tons, equal to over 9 African bush elephants. The idea was to run over the trash that was disgorged from the trash trucks that seemed to arrive at the same time several times a day simply to torment Bob. Bob's interpersonal skills were nil, but that's probably to be expected from a man whose closest personal relationship was with trash.

What's important to understand about a landfill, any landfill, is that air space—the space before the landfill fills up—represents potential money. A lot of money. Conserve air space and money can be made in the future (it should be noted that publicly owned landfills tend to want to conserve landfill space for the future, where private landfills are happy to fill it up as quickly as possible and expand the existing landfill or build a new one). The art and science of compacting trash—moving over the trash to crush the air voids, shred the material, and bind it to other waste—takes a skilled operator to maximize the revenue that can be made in a landfill.

Moving over the waste in different directions helps break the structure of waste down to prevent it from rebounding. The idea is to achieve extreme compaction where densities hover in the 1,700-pounds-per-cubic-yard range. This is equivalent to your 17-gallon kitchen trash bag weighing 143 pounds. Not neutron star-type statistics, where a sugar-cube-size neutron star weighs one hundred million tons, but still impressive.

Air space can also be conserved by minimizing daily cover. Daily cover is, more often than not, sand that is placed on top of the trash at the end of each working day. Federal and state regulations called for a minimum of six inches of daily cover over the entire working face. In our landfill, this meant ten dump-truck loads of sand brought up six days a week. From June 1992, until the landfill closed in July 1999, the same driver delivered every single load. He arrived at the sand pit at 6 AM, parked his pickup truck, loaded his first load, and arrived at the landfill just before 7 AM. Bob would signal to him with a series of wild arm gestures and the more than occasional burst of his air horn to indicate where he wanted Phil to dump his load. I've made up Phil's name, but really, could he be named anything other than Phil? In the seven years I ran the landfill, Phil logged 437,000 miles, enough mileage to circle the earth fifty-five times. He did this in a truck with no air conditioning, no radio, and when I once peered into his truck while he was unlatching his tailgate, I noticed a metal spring piercing the work seat right smack in the middle. A sweet ride this was not. Phil hauled approximately 180,000 tons of daily cover, which not only cost the city over 2.6 million dollars, but also represented 3.1 million dollars in lost air space. Large commercial landfills know exactly how much of a premium air space is, which is why there are alternatives to the sand such as foam sprays or giant tarps that are rolled out and then taken up the next day.

Phil activated the hydraulics which lifted the dump body and when the surface tension between the sand and the metal dump body was broken, the twenty-ton load of sand would slide out, punctuated by the slamming of the tailgate against the dump body. Even when I was expecting the sound, it never failed to startle me. The only equivalent sound that I can think of is a parade featuring Civil War reenactors where they fire off a volley every so often. It's much more of a concussive boom than a crack, and there were many times I was startled silly while working in my office trailer a mere fifty yards from the landfill.

Bob was extremely quirky. He didn't seem to sleep that much, as he had a side scrap business that had him patrolling all over the county in the wee hours of the morning before firing up the compactor at 6 AM. The compactor ran on diesel fuel, Bob ran on coffee. It was some sort of unwritten rule that every single trash-truck driver would bring Bob a cup of coffee on every trip to the landfill. Bob drank upwards of twenty cups of coffee every day, in addition to the thermos of coffee he arrived to work with.

A stout man in his fifties, Bob was very proud of the work he did. And he was good at it. His reputation was that of a crotchety man and his frequent chastising as he bore down on an offender to his carefully constructed working face in the sixty-one-ton compactor with the air horn blaring supported this characterization. I had to calm down many an affronted customer after a run-in with Bob. It might be easy to surmise that this wasn't exactly a challenging job, but there's a fair bit of engineering involved and mixed trash isn't known as the best pyramid building material. Slope and grades had to be precise, and the trash trucks as well as numerous individuals in private vehicles had to be able to get to whatever spot Bob was on, which meant roads and relatively flat spots to dump. And every single day the landscape changed as the landfill rose further and further out of the ground.

Undoubtedly the most difficult waste Bob was forced to deal with was the daily delivery of sludge from the wastewater treatment plant. I don't wear a watch, but every weekday I knew when it was noon, for at that precise time the smell in the lunch room would announce the arrival of the sludge load. If you are not familiar with the wastewater treatment process, one of the first things that happens is the separation of the solids, mostly poo, from the water. Because the solids have so much water content, it has to be run through a press to dry it out, but it doesn't dry to some dusty consistency. It's more like a giant, unstable Jello mold that positively jiggles as it comes off the belt press to splat, the literal sound the sludge makes as it falls, into the dump truck. And at noon we would be treated to the truck passing through the gates of the dump enveloped in a pigpen cloud of freshly pressed poo odor. It's difficult to describe the pungency, but the smell lingered for a good while. The closest approximation is when driving by a dead skunk and tasting its spray as it fills the car with its noxious odor. Those of us in the lunchroom were inconvenienced a bit, but Bob had to work with this stuff. It's hard enough to build a pyramid of trash, but when wet poo is your mortar, it's near impossible. Often the seven-foot-high cleated wheels would get stuck in a deep pocket of sludge and not even the all-wheel drive 560 horsepower machine could extract itself. I'm sure the tow truck drivers that were occasionally dispatched to the landfill dreaded that assignment.

Bob mostly took it in stride. The bane of his existence was when he ran into the perfect storm: a deep pocket of sludge combined with a mattress that unfailingly would wrap its steel coils hopelessly around the axles of the compactor, necessitating removal with a torch to get at the hardened steel coils. I could often find Bob on a coffee or snack break, standing knee-deep in trash mixed with sludge, his bare hands covered in God knows what from unraveling the steel springs entwining the wheels and the only white things visible were his coffee-stained teeth and the bread of the sandwich he was holding.

Bob died on New Year's Day six months before the landfill closed. He pounded over five hundred thousand tons of trash during his decade at the dump. Each afternoon, as I bid Bob farewell and "See you tomorrow" his reply was always the same "Hope so." After only his fifty-sixth year, I never imagined I wouldn't see Bob that next tomorrow. Not many people leave a mountain as their legacy, but when I stand on top of the landfill, the glorious landscape of southwestern New Hampshire unfolding below, I will forever be enjoying the view from Mount Farnsworth.

Chapter 10

My early days as the solid waste manager for the city of Keene were a steep learning curve, particularly when it came to running a landfill. I knew a bit about landfills from my schooling, but actually being responsible for the daily operation of a landfill was real in a way that no book could ever convey. I admit it was often depressing. Yes, the business of waste is about waste, yet it was the magnitude of the waste that was sobering. There were plans to increase recycling, but the fact remained that the vast majority of material arriving at the dump was ultimately going to get buried and this waste of resources bothered me. In my first several weeks, each day after all the workers had gone home I would wander on to the landfill. The daily trash was mostly covered with Phil's ten sand loads, although it was easy to pull a bag out at will to inspect the contents, which I did to reaffirm my commitment to the work I was doing and to keep me humble because without fail every bag I ever tore open was filled with common recyclables. I shared these moments with a lot of seagulls who always kept a wary distance. I also shared these moments with a large flock of turkey vultures. Every Saint Patrick's Day marked the return of these soaring marvels. Shortly after the first frost, they would take flight to warmer climates, but it was uncanny how March 17, without fail, was the day they would return to the landfill. Turkey vultures, also known as turkey buzzards, have a keen sense of smell which can detect the gases produced by decay. Soaring hundreds of feet in the sky with a wingspan that extended up to six feet, this sense of smell, along with veritable eagle eyes, comes in handy for their typical diet of carrion due to roadkill. The landfill provided something akin to an all-you-could-eat buffet. It's pretty eerie to be the lone human amongst these giant birds. Unlike noisy seagulls, the turkey vulture lacks a syrinx—the vocal cord in birds—and the only sounds they make are grunts and hisses. They are, however, menacing sentinels that perched themselves in a row on the concrete walls watching every step I made. I shudder to think what would have happened if I accidentally tripped and knocked myself out. Would it turn into a scene from an Alfred Hitchcock movie? Being picked clean by a swarm of turkey vultures certainly seems like an ignominious ending, although in my compendium of terrible ways to have my life ended, being consumed by turkey vultures is a definite second to being in a fatal car crash involving a trash truck. Happily I've been spared both of these fates.

As if the beady-eyed gaze of dozens of turkey vultures wasn't enough, I really got rattled by a phenomena known as the floating tire. Tires, which are easily recyclable, sometimes find themselves in trash loads and get buried along with everything else. In spite of being crushed by the 120,000-pound compactor, a tire is a tough customer and the structure of the tire often remains intact in spite of the repeated crushing. If a tire does manage to find its way to be buried, it gets exposed to the surrounding biological decomposition process which produces a variety of gases. The lighter gases can build up within the interior of the tire, causing it to act like a balloon that gradually works its way back to the surface. Tires often popped up randomly all over the landfill because of this floating effect.

This process, it turns out, is not confined to tires. I was walking along the dusty road to the top of the landfill my second day on the job when to my horror I saw a baby's arm protruding from the sand. Its hand was clenched in seeming agony. I hurried over to this nightmare, my breathing all but stopped, and I began a gentle excavation around the arm with my bare hands. The arm was attached to a body, and then the head was revealed. A thousand thoughts raced through my brain until it registered that this was a doll. A rubber doll that had floated to the surface. It was the first of many doll arms, heads, or legs that I would see almost daily until the landfill closed. I never got used to it.

The seagulls are long gone now that the landfill has been closed for over two decades, but the turkey vultures remain. They have continued to return to their adopted home because they adapted from the landfill buffet to the transfer station buffet. Now the trash received at the facility is loaded into one-hundred-yard-long open-top trailers for a journey across the state to be buried in a large, corporate-owned landfill. Depending on the volume, up to ten trucks make the ninety-mile, one-way journey. Some of the trucks are loaded in the late afternoon and are covered and staged to be transported the next day. It is these trucks that the turkey vultures feast on. They pull back the mesh covering the trash and excavate their way into the offering until they've had their fill, then they retreat into the nearby woods to await the next day's meal. No doubt there are rubber dolls in many of the twenty-six-ton loads, but they don't have enough time to float to the surface and freak me out. But the turkey vultures do that pretty well all on their own.

Chapter 11

In this day and age, I think it would be considered unusual to have the same boss for over thirty years. The relationship between my boss, Kürt (the umlaut over the u makes the name sound like *Court*, but most people ignore the umlaut and just pronounce it as Kurt, and it happens so often that he doesn't correct anyone anymore), the public works director and me, has lasted longer than my marriage, and when you know someone for that long you inevitably experience the highs and lows of both business and personal matters. If you were to administer a personality profile to each of us, and we've had this done, the results point to very, very different personalities, which we've managed to turn into strengths.

If you knew Kürt (secret tip, depress Alt key and type 0252, and that will place the two dots above the *u* as a shortcut to the umlaut), you would know he's an engineer. If you didn't know him, he'd tell you he's an engineer. In spite of a wide, vast, and deep gulf in how we look at and approach life, I like and respect Kürt. I've often joked that I must be doing something right in his eyes as I haven't had a formal job review in well over a decade, and my code still works on the office door lock, so I keep showing up to work.

One of my value adds is the ability to translate fairly complex ideas into understandable concepts. This is no easy feat when the English language is butchered so badly on a daily basis by the engineers. Kürt is particularly notorious for making a hash of everyday language to the point where my long-term office mate and I created a computer folder that I have titled "Me Talk Good English" to store our collection of "Kürtisms." A small sampling includes:

- "Post humorously" (posthumous)
- "Sympathy orchestra" (symphony orchestra)
- Belly wag or Billy wack (bailiwick)
- "Throw it on the table and see what sticks to the refrigerator" (I think he was going after "throw it against the wall and see what sticks)
- "Suzeami" (tsunami)
- "Tear the baby" (split the baby)
- "Overly autacious" (ostentatious)
- "Girth your loids" (gird your loins)
- "I don't count my chickens before the gun goes off" (don't count your chickens before they hatch)

Most of these jewels occur in our every other week staff meeting, and I've come to look forward to these gathering sessions to add to the collection. Kürt rarely disappoints in this regard.

Kürt came to the city a year after I started, fresh out of graduate school and before that, military service. Kürt served in the US Air Force as a combat engineer, and retired from the military as lieutenant colonel in the US Air Force Reserve. Over the years I've studied Kürt, I've also learned how to think like him. When an exasperated employee would come into my office to ask, "What the hell is Kürt thinking?" I could, with relative ease, put on my Kürt "hat" and explain what I thought his viewpoint was.

Through it all we made a pretty good team. Several years ago, Kürt was named one of the top ten public works directors in the country, which in the public works world is equivalent to winning the Oscar from the Academy of Motion Picture Arts and Sciences. Kürt's award was a validation that our team does good work.

Kürt is a history buff, and, in particular, he knows the history of Keene as few others do. He can literally quote passages from town reports from the 1800s, and his understanding of rules, regulations, codes, statutes, and every other obscure thing is encyclopedic. The kind of knowledge

Kürt retains makes my head hurt. I'm the imaginative, "what-if" thinker of our duo, and to his endless consternation, I still run ideas by him on a weekly basis. I know some make his head hurt. The vigil as to when Kürt will retire had been going on for years, and the answer came at the end of 2023. I will miss Kürt as a boss and a friend. We talk often about the value of the work as an extension of our personas, and we struggle with the dichotomy of having put in countless hours making the trains run on time in our community experiment and of wanting to stay relevant even when we don't occupy these positions of responsibility. For over a decade, I've listened to Kürt talk about starting up a hot dog stand as an antidote to shouldering so much responsibility over so many years. I'll either join him or set up a taco truck across the street where we can yell our memories to each other and carp about how things just "ain't like they used to be." What I do know is there will be many physical legacies of our collaborations left behind, and long after we're gone people in charge will be saying the same thing we often say about our predecessors: "What the hell were they thinking?"

In my three decades in Keene I have worked with four different "CEOs": Pat, John, Med, and Elizabeth. We don't call our chief executive a CEO, we refer to them as the city manager. Keene operates a strong city manager style of governance. Keene has fifteen city councilors: two elected from each of the four voting wards, and seven elected at large, meaning the at-large councilors are elected by the voters of all four wards. An elected mayor rounds out the major elected officials, and the mayor only votes in the event of a tie in the full city council votes.

The city council hires three officials within the government: the city clerk, city attorney, and city manager. The city manager is where the buck stops when it comes to daily operations. I am a direct report to the public works director, and the public works director is a direct report to the city manager. Most every project I work on must be either authorized through our budgeting process, or explicitly authorized by the city council through a vote that then directs the city manager to "get it done." A fifteen-member city council can, at times, be a bit unwieldy, which is where the sub-committees come into play. Each elected councilor is assigned to one of three sub-committees: Finance, Organization, and Personnel (FOP); Municipal Services, Facilities, and Infrastructure (MSFI); and Planning, Lands, and Development (PLD). Any issue that comes before city council is referred to one of the sub-committees, and the sub-committee votes on a recommendation that is forwarded to the full council for a vote. It is very unusual to have a recommendation come out of sub-committee that is not subsequently affirmed by the full city council.

John was the second city manager whom I worked for, and he was a very skilled politician. After a stint in the US Coast Guard, John began working in the municipal sector. His first municipal job was riding on the back of a garbage truck slinging bags into a packer truck. It takes a certain type of personality to aspire to be a city manager, and the job is never boring, but I'd rather have a really bad case of shingles than to ever want to occupy the city manager's chair. I just don't have the temperament, patience, or political skills necessary to succeed in that job, but John sure did. For more than two decades John was Keene's "CEO," and I quickly learned his personality type so as to navigate the political and resource landscape that was either the path or the obstacle to pursue my tree-hugging ways.

I generally worked very well with John. I knew that I had to have my research done and facts in order to bring anything to his attention. Once a project began, it was imperative that everyone be on the same page; even if I was lukewarm on something the city manager wanted, he would cajole me until I affirmed my commitment.

The first time this came to a head was during a review of the financial model used by the Solid Waste Division that I oversaw. Remember, the city's Solid Waste Division was not supported by the tax rate. John always knew he could tweak me by suggesting we could save money by eliminating recycling. I had traveled this road a number of times previously, patiently explaining why it made zero economic sense to do so, and helpfully reminding John that the Solid Waste Fund funneled over five hundred thousand dollars to the general fund to "pay" for the general fund's support of the Solid Waste Division—including the finance department, city attorney, and the city manager. This particular ideological discussion began to get heated as my level of frustration grew from what I felt was a pointless argument. It was right after I pointed out that on a participatory basis recycling is more popular than democracy that John slammed his fist into the conference table and suggested that we should settle our disagreement man to man in the parking lot behind city hall. I'm six feet tall and most days hover around two hundred pounds; John had me by a couple inches and at least fifty pounds, but I reacted immediately, saying, "All right, let's go! Pretty sure I can take you." I was counting on my being fifteen years his junior, my general athleticism, and my rage to be all I needed in my float like a butterfly, sting like a bee plan, although if he managed to get me on the ground things would likely go south quickly.

We both pushed our chairs back in the first choreographed move of being called out when the city attorney who had been watching this escalation with increasing alarm said, "Okay, boys, let's all settle down now." Somehow this intervention broke the spell and tensions soon subsided, but word got around quickly throughout Public Works that I almost got into a fist fight with the city manager, which did wonders for my street cred within the department.

The second confrontation came during one of John's forays into management theory, where we'd be assigned readings from the flavor of the day book such as *Who Moved My Cheese*, *Good to Great*, or my personal favorite, *How to Win Friends and Influence People*. The executive training technique du jour was John getting out of his office and randomly popping by, no specific agenda, just being seen. On this particular day, John came by early in the morning and I had been at my desk for several hours already as I was working on some tight deadlines and I wasn't sure I would complete my assignments on time. I shared an office with the other assistant public works director, Donna. Donna oversaw the water and wastewater utilities in the city. Sharing an office for over ten years, you tend to learn a person's rhythms and habits. Between our desks was a small circular conference table. I knew John was in the building, as I heard him talking to the office manager before I felt his presence enter my office. My back was to the conference table, and I simply kept furiously pecking away at my computer (kind of like I'm doing now). It is true that I was up against a deadline, but I was also deep into the unique low-vibration experience that so many of you have likely experienced—divorce. I was stressed about my deadline, in a foul mood because my divorce, and I just didn't have time to make nice when John broke the silence with a long, drawn out "Gooooooooood morning." I wheeled my chair around, said my own exaggerated "Gooooooood morning," and spun my chair back to my computer to resume typing. John put on his best parochial voice and began saying, "I can tell that you're feeling . . ." when I whipped around, leaned in pointing a finger at his face, and said, "You don't get to tell me how I'm feeling!"

Donna rescued me big time, as she could see how this was going to spiral quickly out of control, knowing our personalities. Donna hustled John out of the office under the guise of wanting to show him something in the map room. The fallout was fairly swift. Later that day, the public works director informed me that I was disinvited from our weekly senior staff meeting up at city hall. Really? This felt on par with Dean Wormer's "double secret probation" penalty from *Animal House*. I became the envy of all those that still had to go to the staff meetings. About two years later, I was finally allowed to resume attending.

And John, if you happen to be reading this, I think you were a helluva city manager and I learned a lot from you, but there's not a shred of doubt in my mind—I definitely would have taken you that day in the parking lot (wink, wink).

Chapter 12

It was in the near-term aftermath of 9/11, my mind mostly numb from the sheer horror of the carnage and destruction, that I reflected on the landfill with a different perspective. The glorious weather that dawned in New England on 9/11 had me driving to a workshop on dealing with asbestos waste. I was just pulling into the venue when the first news reports started to come over the radio. I immediately called my brother, who lives in Manhattan, because I wasn't sure if he was in proximity of the World Trade Center, in the confusion it seemed the whole island was under siege, and I just wanted to hear his voice. It wasn't a surprise, but it was very unsettling to hear the "All circuits are busy" message from the attempted cell phone call. I entered the workshop feeling very uneasy.

A television had been turned on in the workshop room and the visual images began to complement the audio descriptions I had heard on the radio. Then one of the towers collapsed and it quickly became apparent that the world had forever changed. I wondered how I would explain this to my children, then aged four and one. Explaining evil of that magnitude would have to wait for later years. I needed to focus on something familiar and in the days following the attack I began to contemplate the logistics of the cleanup that would unfold in the following months and wrote an article for a local weekly *Shopper News* paper that circulates in the county as follows.

Don't Recycle Hate

The horror of the terrorist attacks at the World Trade Center, the Pentagon and in rural Pennsylvania are beyond my ability to articulate. In many ways denial allows us to cope with overwhelming circumstances. A few days after the attacks, I started to think about the waste management problem that would unfold in the next few months and it was truly mind numbing.

A cubic yard of concrete weighs around 4,000 pounds. The Twin Towers were constructed with over 425,000 cubic yards of concrete. Even with the biggest trucks it would take more than thirty thousand trucks to remove the concrete alone. There was also more than 200,000 tons of steel which would require another seven thousand trucks to haul away. The rubble approached two million tons which is 120% more than what is contained in the closed Keene landfill. What took the Keene area over forty years to accumulate in a landfill was exceeded in mere hours. Being in an asbestos workshop at the time I wondered if the cloud of ash that seemed to hover for days was laden with asbestos. After doing a bit of research I discovered that the World Trade Center was built without asbestos and the air monitoring conducted by the Environmental Protection Agency detected no elevated levels of asbestos. It's been well documented that the health of first responders and cleanup crews working in and around ground zero have been severely compromised as there was undoubtedly a toxic stew emanating from the smoldering rubble. It's only a token bit of good news that the people caught in the ash cloud and forced to breath the chocking dust did not add the pernicious effects of asbestosis as a cruel whammy to the already mind-numbing tragedy.

The 200,000 tons of steel has been recycled, or reused through displays in all fifty states of the twisted metal serving as a reminder of that infamous day. Fresh Kills landfill, the notorious repository located in Staten Island, which is home to 150 million tons of New York City Waste buried throughout the former estuary, was reopened to deal with the World Trade Center debris. While many personal effects were recovered at ground zero, over 1,600 personal items were sorted at Fresh Kills further magnifying the countless personal tragedies of the attack on the nation. The Pentagon, which was also attacked, generated its own 15,000 tons of debris. We buried the debris, and in spite of the devastation these truths are evident—The American spirit cannot be buried and hate should not be recycled.

The Freedom Towers were erected from the ruins of the World Trade Center site, the Pentagon was rebuilt, and the Pennsylvania field where the heroic passengers of Flight 93 prevented a greater tragedy now has a memorial surrounded by trees and shrubs, life springing from the ashes It was interesting that after that article was published I got messages from as far away as Washington State supporting the notion that we may bend, but we will not break in the face of tyranny.

It was interesting that after that article was published I got messages from as far away as Washington State supporting the notion that we may bend, but we will not break in the face of tyranny.

The author joined the Public Works Department in Keene, NH, in 1992. Pictured here in front of bale of HDPE plastic.

Aerial view of facility. Closed landfill on right.

The author in 1992. "Recycling saves you money. It really does!"

Keene Recycling Center in 1992.

Keene Transfer Station and Recycling Center in 2023.

Northeast Resource Recovery Association's "Recycler of the Year Award" in 1995.

Partial Crew in 1999 (from left to right): Hank Capellano, Al Fisk, Allen Austin, Cliff LaFleur, Duncan Watson, Fred Hale, Bob Paddock.

The author during the Charlie Brown years.

The author in 2024, in front of a new bale of HDPE plastic.

Sunrise at the dump. Mt. Monadnock in the background. Photo by Fred Hale.

ARTIST AT LARGE

RECYCLING IN HELL
by Roz Chast

Close to reality.

Chapter 13

I am a person who holds an unshakeable faith in the fundamental goodness of the vast majority of humans. Spending time at the dump can reaffirm this or further convince me that the inmates run the asylum. It is a working theory that a surprising number of people believe that magic fairies and elves ply the recycling center at night to fix the quality-control issues that vex us.

I won't soon forget the woman who was dropping off her glass at the recycling center. She was dutifully putting the separated glass in the bin when she pulled out a clear bottle with a dead mouse inside it. As she tossed the bottle in with the other recyclables she casually remarked that it would be taken care of when the glass was melted. I'm not sure what miracle process she's referring to, but a dead mouse is most definitely a contaminant which impacts our ability to provide the raw materials that can be made into new products. Without consistent quality, no producer will buy our goods, and then there are no options but to dispose of the material as trash. I'm going to guess that the woman didn't put the mouse in the bottle. The mouse climbed in searching for food, got stuck, and died. Things like this happen, but now we have a potential proverbial bad apple that could spoil the barrel. If you go to your favorite restaurant to get a sandwich, and there is some chopped up mouse mixed in with your chicken salad is it okay for that to be in there even though your stomach will likely be able to digest it? Of course not. The glass mill wants glass, not glass with leftover spaghetti sauce, peanut butter, mayonnaise, or dead mice.

Our asylum is fairly spread out. When entering the dump, the first stop is the scale house where the preternaturally friendly attendants will assess what the customer has and direct them to the right place or places on the site. Of course, the most common items are trash and recycling, but we also have areas for scrap metal, tires, brush, compost, household hazardous waste, cooking oil, electronics, appliances that contain refrigerants, automobile batteries, textiles, shingles, and construction debris. With a total staff at any given time of nine, seven of them either stationed inside the recycling center or at the transfer station, it's impossible to be in all places at all times, which results in daily opportunities for our customers to run amok.

A typical week would be as follows:

On Monday while walking the site, I came across a pile of wood waste and assorted refuse, including a partially full can of insecticide, that someone deposited on our active compost pile. What was ironic about it was that the material was placed in and around the sign that explained what can and cannot be placed in the pile.

On Tuesday, we discovered missing rip rap. Missing rip rap? The closed landfill is ringed with drainage swales. The purpose of the drainage swales is to channel precipitation in the form of rain or melting snow in a controlled fashion. The swales are lined with rough crushed stone six-to-nine inches in size. The stone is referred to as rip rap. The rip rap helps prevent erosion in the drainage swale by slowing the velocity of water in a heavy precipitation event as well as stabilizing the ground underneath. It's not a particularly valuable landscaping product, so why someone would pick a spot away from all the possible sight lines of our personnel and then lug a six-foot-wide by twenty-foot-long section of rip rap to their pickup truck which, because of the terrain, couldn't have been parked closer than fifty feet from the swale? The audacity and determination of the person who undertook this theft was impressive, but on Wednesday it was rivaled.

Typically, Wednesday is a sucking day. Appliances that contain refrigerants have freon that must be removed and recycled before the appliance can be sent to the scrap metal recycler. Freon, when released into the atmosphere, contributes to the deterioration of the ozone layer that, in part, makes our planet habitable, so protecting the ozone layer is a good idea. The machine we use to remove the refrigerants is run by a small generator. On sucking day, the machine is set in place; a vacuum tube is attached to the refrigerant line of a freezer, refrigerator, or air conditioner; the generator is engaged; the sucking of the freon begins; and fifteen minutes later a staff member returns to hook up a new batch. Up to six appliances can be evacuated at any one time, and rather than stand idly by while the sucker does its work, another chore is undertaken. On this Wednesday, the employee returned to disconnect the sucker only to find all of the vacuum tubes neatly coiled, including the extension cord that ran from the generator to the sucker, but no generator. Right from under our noses the generator was stolen. Maybe the rationale was "Hey look, here's a perfectly good generator next to the scrap metal pile. It's running and seems to be attached to all these refrigerators, but it's near the scrap metal pile, it isn't nailed down, therefore no one wants it."

Thursday is the day the yard sale professionals show arrives. For years, the dump had a swap shop or salvage barn. The concept was simple—if you had something that you no longer wanted, was in good condition, and perhaps someone else could use it, then it could be deposited at the swap shop free of charge. Mostly the salvage barn made a mockery of the saying one person's trash is another person's treasure. It was much more like one person's trash is another person's trash, or in our case, our trash. The swap shop had to be constantly monitored to ensure people weren't using it to avoid paying to dispose of the item they claimed was a treasure. It is always interesting to catch someone in the act of clearly violating the spirit of the salvage barn. The woman putting in the ceramic lamp that looks like someone took a hammer to it in a fit of rage, small pieces of it hanging off the frame with jagged edges, the wiring exposed, and of questionable safety. The man pulling up in his Mercedes to drop off the waffle iron with the power cord frayed to the point where it was completely unusable. I was certain he had run over the cord with his lawn mower, but that didn't deter him. It's like the people who deposit things in the swap shop leave their shame at home. Dozens of these items arrived each day. Maybe 10 percent of the items were still serviceable in some way, but it wasn't like we had a workshop to fix the mostly broken items we received. Sure, many of the items could be fixed. The waffle iron, for instance, could have had a new power cord installed. But in addition to replacing the frayed power cord it had years of dried-up batter caked on the sides and it didn't look like something you'd want to take home to prepare Sunday waffle breakfast, particularly when you could go to the local department store and buy a brand-new waffle iron for $15, maybe $10 if it was on sale. No, this stuff was mostly trash, and the people bringing it to us were either naively optimistic about the repair abilities of the people perusing the swap shop, or, more likely, they had something they no longer wanted because it was broken and they just didn't want to pay to get rid of it. To be fair, there was the odd treasure, something that was actually in good condition, and for whatever reason it was no longer wanted or needed. The bathtub in my house is a testament to this notion. I just happened to be lucky enough to be at the scale house when a pickup truck with a brand-new bathtub in the bed of the truck pulled in. The story was the woman doing a home renovation had ordered a new tub and this one showed up, but it wasn't the color she wanted, so she told the contractor to get rid of it. I had the pickup truck back up to my station wagon and with more than a little Yankee ingenuity managed to get this three-hundred-pound cast-iron tub off the pickup truck and into my wagon. A year later, I had it installed in my own home renovation, color be damned. I saved $600 and all it cost me was a check to the city for the value of the scrap metal that the city would have realized if it had been sold as scrap (if my memory serves, it was under $5).

It was the professional yard sellers that ultimately killed the swap shop, although it died a slow death. Thursdays, the pros—the people who had yard sales every single weekend during the warmer months—would swoop in like seagulls converging on a French fry at the beach and they would pick the swap shop clean. No item would be left, whether coffee mugs, soccer cleats, worn-out dining chairs, or questionable crock pots. Over the course of Thursday morning, a half-dozen mini vans would arrive at the dump and tell the attendant they were headed to the swap shop. At first I thought this was pretty great. Reuse is a higher and better use, and if someone made a few dollars off of it I didn't much care. And then the flaw became apparent. As I said before, maybe 10 percent of the material that came to the salvage barn had any real value as an item of reuse. The professional yard sale scavengers would take anything that wasn't nailed down and try to sell it. Their pricing structure was on the order of ten cents to five dollars depending on the item. The coffee mug with a picture of a third-grade student that was probably given as a gift to a teacher or a relative—that was a ten-cent item. Any item they took had a four-week window of opportunity to sell. And inevitably when the picture coffee mug and hundreds of other similar items failed to fetch a dime, they ended up back with us. If something isn't worth ten cents to buy, it isn't going to go for free either. This is how we ended up with 90 percent of the material back in the salvage barn. We had, of course, seen this stuff before, so now it was up to us to deliver it to its final resting place—where it probably belonged in the first place.

In general I'm a huge fan of reuse. Kudos to Goodwill, Salvation Army, and hundreds of church-run thrift shops where legitimately good items are given new life for pennies on the dollar to the purchaser. Our experiment with a swap shop was revealed to be a way for people to avoid paying for something they wanted to get rid of. Never underestimate the lengths people will go to save a nickel.

Friday is nickel day. Nickels refer to redemption containers that are sold in bottle bill states. Currently the states of California, Connecticut,

Delaware, Hawaii, Iowa, Maine, Massachusetts, Michigan, New York, Oregon, and Vermont have bottle bills. New Hampshire is literally surrounded by bottle bill states, but New Hampshire does not have a bottle bill in spite of bottle bill legislation being proposed many, many, many times. New Hampshire and Rhode Island are the lone New England states without a bottle bill, and if you ever want to raise the ire of anyone that is involved in recycling in New Hampshire, the Live Free or Die State, just ask for an opinion about bringing a bottle bill to New Hampshire. Passions are high for and against. New Hampshire, supported by a well-organized grocery association, has long argued it's a tax and an unnecessary one at that because the grocers association fund a volunteer-run roadside cleanup system run by New Hampshire the Beautiful. Companies and civic organizations adopt a section of state highways and New Hampshire the Beautiful provides the bags for roadside cleanup and the signage that gives the company or civic organization good public relations visibility. And for the most part it works. Bottle bills are proven to reduce roadside litter by creating an economic incentive to collect the beverage container, be it a beer can or bottle, soda can or bottle, and states such as Maine that have expanded the types of containers eligible for redemption.

The way it's supposed to work is the beverage distributor is only supposed to distribute containers marked for redemption in states that have bottle bills. With four of the six New England states offering redemption of containers, the beverage manufacturers certainly don't do a special production run for New Hampshire or Rhode Island. No, they simply ship redemption containers marked for redemption to our non-redemption state. While there's no telling whether the redemption containers we see at the recycling center were purchased in nearby Massachusetts or Vermont, with the corresponding nickel that is supposed to be collected in the redemption state to fund the program, what we do know is that we now have a container that is worth five cents a mere twenty miles away. On our sorting line, we segregate these redemption containers.

These nickels add up. We collect tens of thousands of dollars a year, which helps fund our recycling efforts. It's a little less of a problem in Vermont, where the redemption rate is in the 80 percent range, which means 80 percent of all the containers sold in Vermont for which a nickel was collected when the product was sold are returned. This 80 percent redemption rate includes every single New Hampshire border community doing exactly what we do—returning these containers that certainly could have been sold in Vermont, but somehow ended up at the Keene Recycling Center or the Walpole Recycling Center or the Claremont Recycling Center, working its way right up to Canada.

The simple fact is that it's illegal to distribute a redemption container in a non-redemption state, but when something like an aluminum can becomes seventeen times more valuable as a nickel return than baling it up and sending it to the aluminum smelter, it's just too good an economic opportunity to pass up. And that nickel container will eventually end up at the aluminum smelter anyway.

In Maine, it's a slightly different story. Maine has an expanded bottle bill where more containers are worth a nickel. The big bonanza here is water bottles. In spite of active campaigns to reduce the consumption of plastic water bottles, in 2022 over 50 BILLION with a B were sold in the United States. Maine doesn't quite match the incentive offered by the state of Michigan, where a redemption container is worth ten cents, but think about how many plastic water bottles you see on an average day and you can understand how the numbers add up. Schemes to take advantage of redemption containers abound. Popular culture highlighted the incentive in an infamous *Seinfeld* episode where Kramer and Newman loaded an empty US mail tractor trailer that was deadheading (being transported empty), with redemption containers collected in New York City where they were worth five cents destined for Michigan where they were worth double in an attempt to game the system. In Maine, the redemption rate *exceeds* 100 percent, which makes the state take greater notice to ensure that only Maine containers are returned to Maine. It's a true cat-and-mouse game with high stakes. The estimated total value of redemption containers sold in Maine is $36 million, and in Vermont it's $600,000. Just doing some quick math taking the population of Maine and Vermont to get a per capita rate of redemption containers and assuming the rate of redemption containers sold in New Hampshire to be the same order of magnitude as Maine and Vermont, the value of containers marked for redemption in New Hampshire is $36 million if redeemed in Vermont or Maine. Of course, nowhere near that level of redemption of New Hampshire containers occurs in either Vermont or Maine, but that kind of money creates incentive.

There is no delicate way to put this. Saturday is a trainwreck. During the planning for the transition from the landfill to the recycling center and transfer station, it was not anticipated how popular the place would become. Word got around how great the place is, but what drives a lot of people to the dump is the money saved by hauling your own stuff versus a private contractor to pick up the stuff at curbside. While the vast majority of Keene residents opt for the convenience of curbside collection, there are increasing numbers of people, particularly on Saturdays, that pay us a visit.

The solid waste system in Keene is operated as a special revenue fund, which means that property taxes do not pay for collection and disposal of refuse or recycling. New Hampshire has no sales or income tax, but boy do we have property taxes. Most communities in New Hampshire build the cost of their solid waste operations into the property tax rate, but not Keene. We live and die by the fees collected for refuse disposal, also known as tipping fees, as well as the revenue generated from the sale of processed recyclables. In our city of 23,000, our business generates about $5 million in revenue and $5 million in expenses. If refuse and recycling is collected at curbside, which about 90 percent of Keene residents do, the cost is around $40–50 per month for weekly collection. If you decide to bring up your own stuff, each bag of trash costs $2, and recycling is free. A few years ago, I got a letter from a resident that illustrates the potential. It read:

Dear Mr. Watson,
Last Friday I took thirteen or fourteen plastic bags of material to the landfill. I stopped first at the recycling center to unload the cans, bottles, papers, magazines and cardboard. After I had done that, to my great surprise, I had only three bags of refuse to go to the landfill. This meant that over 75% of my refuse was recyclable. I was surprised at this because I had made only a perfunctory effort at recycling and I knew at least 50% of the three bags that went to the landfill had recyclable paper. I mentioned this to a friend of mine. He was not surprised. He replied that he recycles about 90% of his refuse. He goes to the recycling center once a month and to the landfill once a year.

Letters like this give me hope that the throwaway society that we have developed over several generations can be changed to view refuse as a resource that has an economic and environmental value. Know that your actions do make a difference. To hammer this point, as I wrote this, I was sitting on the porch at my sister-in-law's house while vacationing in Atlantic City, NJ. The house next door had their trash barrel at the curb loaded with stuff for the landfill, it just hadn't been picked up yet. Next to the barrel was a set of patio furniture. It's hard to tell if it is made of steel or aluminum, but it had seen better days and was discarded by its owners. Like the proverbial butterfly landing on my finger to remind me of the glory, mystery, and beauty of nature, a beater pickup truck came down the street looking for things like the patio furniture. The pickup truck was already piled high with scrap metal and it took only a few minutes before he had added the patio furniture to his jigsaw puzzle. And just like that he was gone and so was the patio furniture. It was a beautiful thing to witness.

Saturdays are what I refer to as amateur hour. The trip to the dump on a Saturday is a ritual. And it doesn't matter what the weather is doing. We have had some epic blizzards, but try to stop the men (it's always the men) from making their trip to the dump to get rid of their recycling and you might have a mutiny on your hands. Most Saturdays are just crazy busy. The lineup of cars can extend a quarter-mile down the road, but mostly people are good natured. There's almost always a great view, and it's a place to catch up with your friends and neighbors, and feel like you're doing something to cross a chore off the list or just maybe doing a small thing to help protect the environment.

Sunday we are closed, but the dump is never far from my mind. I'll take Sunday to think about challenges such as how to manage the disposal of products such as the sneaker that lights up every time you take a step. It was, at the time, a brilliant marketing idea. In the competitive world of athletic shoes, a new novelty can mean tens of millions in revenue. Seven million pairs of the new LA Gear light-up shoes were sold in a few short years at $50–70 a pair. The problem was with the design of the electrical switch that turned the light on and off. The switch contained mercury, which if you don't already know, is a hazardous waste that if somehow ingested or inhaled can create some very unfortunate health effects. When the shoe wore out, and because they were pretty cool they wore out rather quickly, the shoe ended up in the landfill or incinerator. After a series of fines and threatened lawsuits, LA Gear replaced the mercury switch with a ball bearing design. LA Gear never anticipated a problem with disposal (I'll give them the benefit of the doubt rather than a callous indifference to the environmental consequences of their business decision to use mercury), and to their credit they developed a take-back program to produce a responsible disposal option. But in the meantime, how many millions of these shoes ended up at dumps like mine? The geomorphology of Keene's unlined landfill is about as good as an unlined landfill can get. Under all that trash is a thick layer of glacial till which does not have many pores to allow the movement of water. Anything that could be a possible contaminant will eventually make its way into surrounding groundwater, and even though will happen at glacial speed, it doesn't mean it's okay.

While I question the social need of a shoe that lights up each time a step is taken, it's just one example of a long line of marketing decisions that influence consumer behavior without regard for the environmental consequences of those decisions. And those of us that work at the dump are again put in the position of dealing with a problem at the end of the pipe that originated at the beginning. Wouldn't it be something if at marketing meetings, where these types of products are conceived, there were end of the pipe experts that brought an awareness of the product's final disposition? There are a few out there, but we need more.

Chapter 14

Every day of work is ripe with the foibles of life. The dump is a great equalizer, and if you choose to save a few dollars by bringing your own things for disposal you'll wait in the same line and be directed to the same place as everyone else. It's not easy to keep dump access roads clear, but they were always reasonably delineated and we took care to make the experience as friendly as possible. We had a regular visitor in a former mayor who had a local talk radio show that mostly featured his conservative leanings. He frequently directed his considerable ire of the grossly inefficient dump operation and the hack (me), who was running the show as emblematic of everything that was wrong with the world. From my observation, his primary personality trait was impatience, and one day he arrived at the dump when we had an unusual backup of cars. It was either just a busy time or something operationally was holding up the line. Regardless, he queued into the line in his shiny, black, late-model, four-wheel-drive pickup truck with chrome rims and he actually started to honk his horn as if somehow this would get people moving. After repeatedly honking, he pulled into the open lane and sped past the cars waiting in line, past the scale house where the attendant barely had time to raise his palm to direct him to stop and, into the landfill he went. There was a small crowd of us conversing about some operational issue, and we all turned to watch the former mayor disregard the access road and begin to climb the active face of the landfill. This was a 2-to-1 slope of partially compacted trash. The only thing capable of getting up that slope is the compactor itself—the 560-horsepower, 61-ton, cleated-wheel, all-wheel-drive machine, not the 250-horsepower, 4,000-pound, four-wheel-drive Ford F-150. It was, to be honest, a valiant attempt, though. He came to a stop about halfway up the slope, and we figured that he was just then putting it into four-wheel drive because he started to inch forward, wheels digging into the trash in search of traction. The engine was revving very high, he was again making some progress, and he might have made it if he hadn't hit that pocket of sludge. A reasonably fresh load of sludge had been delivered earlier that day (at the stroke of noon as always), and Bob always covered it over with a bit of trash to keep the smell down a bit. The former mayor sank well below his axles and no amount of acceleration was going to help him now. The churning action of the wheels in the sludge was creating a very ripe environment. The more frustrated he got, the more he depressed the gas pedal and now the truck, in its search for traction, was spinning all four wheels and flinging human feces throughout the undercarriage of his truck. And then all was calm. The truck was still running, but he must have been reassessing his options. This spectacle of the attempted hill climb had attracted quite the throng of amused observers over his five-minute attempt. It was now time to raise the white flag. I saw the reverse lights come on and he began his descent. This is the other thing about a hill made of trash—it's not conducive to vehicular brakes. The brake lights lit up, but he was still picking up speed when he finally made it back to level ground, and he just didn't have enough room before he went up on one of our make-shift loading docks, only stopping when the undercarriage of the truck struck the ground because the rear half of the truck was suspended in the air hanging off the end of the loading dock.

    In fairly short order the guys hooked a chain on to his tow bar, attached the other end to the loader, and we pulled him off the loading dock. Without so much as a thank you he drove off without even depositing the four bags of trash he had brought in the first place. A week later, I received a bill from the former mayor for $800—the cost of replacing the ruined tires on his truck. I was outraged by the gall of this guy sending a bill, but people much more politically savvy than me decided it was far more expedient to pay the bill through the city's insurance company and move on with life rather than escalating the situation. He never mentioned the dump or the hack manager on his radio show again.

Chapter 15

Last week (relatively speaking), after over two decades in the scale house, Diane retired. Every Thursday and Friday from 7 AM to 3 PM and Saturdays from 8 AM to 1 PM, Diane was the first face you saw when visiting us. It takes a unique skill set to deal with the variety of humanity that routinely comes to the dump, and Diane was simply brilliant. I'm convinced she could have successfully run for mayor and won, and her renowned popularity literally opened doors that would have been closed to most people. In Diane's off hours, she would go door to door evangelizing her Jehovah's Witness faith. More often than not when the person answering the door realized it was the dump lady at the door they would let her in to speak her piece. I have no idea if she's ever been successful at converting anyone to her faith, but she has touched a lot of lives in her twenty-plus years as the heart and soul of our operation. To meet Diane, is to be surprised that such a larger-than-life persona can be packed into her 5-foot, 2-inch dynamo frame.

Diane had a saying: "You can't save the world, some people don't recycle, and others lie." But she patiently answered thousands upon thousands of questions either via phone or in person with the same cheerful tone. She was called the dump lady, dump diva, sweetheart, honey, and plenty of "gross words" that she preferred to keep to herself. If the question wasn't about recycling or how to get rid of something, she was often asked if her hair was real, how often she washed it, was she married, how could she eat her lunch with how bad the place smelled, was there heat in the scale house, was there air conditioning in the scale house, did the city buy the dog biscuits she handed out? Diane was asked out numerous times, including one proposal with the prelude that his wife had died the past week and might she be available for supper the following week after the funeral.

Whenever anyone saw her in a place other than the scale house, they would almost always remark that she seemed much shorter than they imagined. Short as she was, she had a good view from the scale house window where she saw too much skin on both men and women, too much cleavage on both men and women, no underwear on both men and women, pajamas of all stripes, people wearing flip flops in the last place you'd ever want to have your feet exposed, and more than her share of extremely sweaty people.

She also saw young children riding in the front seat without seat belts, dogs drooling on their owner's shoulder, people holding their punch cards in their mouth. When that happened, Diane would hand over the hole puncher for them to do the job or get a free lip piercing. A woman who liked to take her parrot for a ride, or the dog who jumped out of the back of the pickup truck unbeknownst to the owner. Diane kept him safe in the scale house and the owner had to eventually be called at home to come pick up his dog. Dogs eagerly anticipated the trip to the dump because Pavlov was always ready to dole out a treat. Many a dog owner would break down in tears telling Diane of their dogs passing, thanking Diane for all the biscuits she provided, and paying it forward with a box of biscuits for all the dogs to come.

Diane started working at the dump when we had an active landfill. Previously, I mentioned the seagulls that took up residence at the dump because we had what amounted to an all-you-can-eat buffet. Diane started to learn about the seagulls behavior from her eyewitness observations. She augmented that knowledge when the US Fish and Wildlife Service showed up to take a seagull census. It was then that we all learned that our fine feathered friends numbered in the thousands by official count. At the slightest provocation, the seagulls would take flight, following what seemed like a choreographed flight path before landing on the roof of the recycling center until it seemed safe to return to feeding on the landfill. Their flight path took them directly over the scale house where Diane knowingly took refuge because in spite of the cloudless skies, when the seagulls were airborne rain of a different sort was sure to follow. Right next to the scale house was the swap shop where people could bring in items in good repair to find a new home. One particular day, there was a woman impeccably dressed in a beautiful designer red coat perusing the swap shop hoping to find a treasure. Nothing unusual in that, but as she watched the woman looking over the items, she heard Bob start up the landfill compactor, which meant in an instant the sky would be full of seagulls following their preferred flight path over the scale house and the swap shop. Diane scrambled to the door, opened it and yelled to the woman, "Stay where you are! Don't move!" Diane meant for the woman to stay under the roof of the swap shop, but she either didn't hear or didn't understand, because she stepped out from under cover, making her way to her car at the exact moment the seagulls appeared overhead. Within seconds, the woman's coat turned from bright red to a Jackson Pollock painting of white and yellow with traces of red. She must have been hit thirty times by what is a surprising amount of bird shit in terms of volume from a single bird. She wasn't the only person to be so unfortunately festooned, but she was certainly the most memorable.

Death is the bargain we all sign up for as a condition of life, and I've reached that point in my career where family members of colleagues or former colleagues reach the end of the line. I was reeling with the recent loss of my father when I learned that Diane's father had recently passed. Long before Diane retired and through the first months of her retirement, Diane had been the primary caregiver to her father, and I knew they were very close. When I reached Diane via phone to express my condolences, we exchanged stories of the roles these important men—our fathers—had played in our lives. As we were wrapping up the call, Diane said, "Oh by the way, I have a good story for you."

Several years prior, Diane was manning the scale house when a pickup truck from one of the local monument companies arrived with several ground marker gravestones in the back. Diane asked what they intended to do with the markers and was told that the markers were misspellings and they were going to put them in the rubble pile. Diane waved them over to her truck and took the markers, thinking she would use them as decorative steps in her garden or some other undesignated use. After her father's death, Diane reached out to a friend who had the skills to polish and engrave the stones, so she had one of the misspelled markers turned over, polished, and engraved with her father's name. Diane saved a bunch of money and perhaps is in the running for most creative reuse of an item that otherwise would have been turned into something of far less value. As she delivered the newly polished and engraved marker to the cemetery, she left the groundskeeper with very specific instruction: "Just make sure that 'Hamilten' (originally was supposed to be 'Hamilton'), is facing down and 'Castor' is facing up." I got a good belly laugh out of that one. Classic Diane.

It was the story Diane told me about George that was the most memorable for me. I'm not sure how it came to be that George and Diane started a four-year relationship, but likely it was a phone call that first connected them. The first call came at 7:30 AM on a Thursday, a half hour into Diane's shift. And every Thursday for four years at precisely 7:30 AM, George called Diane and went through his script, never varying. "Constipation, what's it for? Makes me nervous. Hurts, hurts. Prune juice, what's it for?" Most people, I imagine, would not be as indulgent to something like this, but Diane accepted the ritual in the kindest way, giving an ear to someone who needed to be heard.

Whether George was obsessive compulsive or had some other brain disorder is best left for professionals to diagnose, but I was fascinated by the longevity of the repetitive pattern. Years before I found my way to the dump, when I worked in my family's restaurant, Lou's in Hanover, NH, every Sunday through high school I was a waiter in a 1947-style diner with Formica counters and swivel stools. The pace was fast, the food tasty and hot. It was a community go-to place, especially on a Sunday. I like the learning process of becoming efficient, probably because of my own obsessive-compulsive tendencies. My thing was no wasted motions. If you had time to lean, you had time to clean, and whether I was bringing stuff out of the kitchen or into the kitchen, I was plotting my next move for the sake of efficiency. Of course, I tried to humanize the experience because the characters who occupied my stools were interesting in unexpected ways. The counter of the restaurant was a great equalizer. First come, first serve, and whether you were the school custodian or the flavor of the day politician (New Hampshire, and this restaurant, were must-stops for anyone aspiring to be President), you were treated the same and served in order. Rumor had it that we had good coffee. It sure smelled good, but I never acquired a taste for it. But people came for the coffee as much as the good food. This is how I came to meet Jimmy. I don't actually know his name, but I'd be surprised if it wasn't Jimmy. Not Jim or James. Jimmy. Facing me along the row of about fourteen stools, Jimmy had to sit on my far right. That first stool was separated by a takeout counter and ensured that only one person could sit next to him, to his right. His first order was a ham and cheese omelet, dry wheat toast, and black coffee. Not unusual by any stretch, but a pattern emerged as every single Sunday at exactly the same time Jimmy would appear, wait for the stool of choice to open up, set his *Boston Globe* on the counter, pick up a menu, study it, put it down, and wait expectantly for me to approach him to take his order. Without making eye contact he would order his ham and cheese omelet, dry wheat toast, and black coffee. About six months into this pattern, I decided to anticipate his arrival. I spied him at the precise moment of time going into the convenience store across the street to purchase the *Boston Globe* before coming into the restaurant. I had three minutes to complete my mission. I sprinted to the kitchen and ordered a ham and cheese omelet, dropped the wheat toast into the toaster, and poured a cup of coffee. The restaurant was a bit slower than usual that Sunday and Jimmy's preferred counter seat was open. Immediately upon sitting down, I ceremoniously placed a ham and cheese omelet, dry wheat toast, and a cup of black coffee in front of him. I was extremely pleased with myself, knowing my customers so well that I filled his order without him even having to tell me what it was. But his reaction was not what I expected. "What is this?" he asked as if startled. I explained the obvious: "It's your ham and cheese omelet, dry wheat toast, and black coffee" I said cheerfully. That's when he switched from startled to affronted: "I did not order this." And while it was true he had not ordered what was in front of him, he had ordered this exact thing every single Sunday for the past six months. What I thought was good customer service was actually an unacceptable break in his routine. *His* routine, not my routine. At that age, I wasn't savvy enough to understand what was going on, but Jimmy left the restaurant that day never to return. And like George, I'll always wonder what became of them. There is no question that Diane's life was made richer by knowing George, just like my life was enriched for knowing Jimmy. A little compassion and a little empathy go a long way in this world.

Sometimes compassion and empathy isn't enough. At one point after the landfill closed and the transfer station opened, the city subcontracted out the operation of the transfer station to a behemoth company that has the words "waste" and "management" as part of the company name. A transfer station is typically a large metal building with concrete walls on two sides of the building extending up about fifteen feet. The metal building sits on top of the concrete walls, creating a ceiling height of about thirty feet. Refuse trucks that collect trash around town arrive at the transfer station to offload their material. A modern compactor truck can hold around ten tons of refuse. After the contents of the packer truck are disgorged onto the reinforced concrete floor, a loader pushes the material in towards the concrete walls that serve as a push wall where the trash can be piled high and to make room for the next compactor truck to unload.

Picking away at this mound of trash in the fifty-to-sixty-ton range is an excavator that is perched against a knee-high wall that separates the upper level of the transfer station from the lower level. In the lower level of the transfer station, which drops about fifteen feet from the upper level, known as the tunnel, is a one-hundred-yard, aluminum-bodied, open-top trailer. The job of the excavator operator is that of a puzzle master. Plucking the trash from the mound against the push wall with a grapple arm (hundreds of pounds are gathered in one scoop), the excavator swivels, and the material is placed into the open top trailer. The skill comes in the actual placement of the material because the trailer needs to be balanced and space needs to be optimized to come as close to the maximum legal weight load (around twenty-six tons) as possible. This balance and space optimization is harder than it would seem to be, because the material being loaded is never consistent. Picking into the pile on the tip floor is just as likely to result in a bucket full of concrete rubble as it is to dig into large Styrofoam panels that, while bulky, weigh next to nothing. Everything in a transfer station is about minimizing the cost of transportation. If a transfer station of even moderate size—and the Keene Transfer Station is definitely modestly sized, at about thirty thousand tons shipped annually—ships each trailer at the maximum weight versus, say, two tons lighter, that differential weight will cost an additional $50,000 in transport costs per year.

When the transfer station is open, there are a pair of twenty-four-foot rolling doors that are open to the elements, particularly wind. Because of the frequency of trash trucks visiting the transfer station (30,000 tons annually, an average refuse truck contains 10 tons, this equates to 3,000 trucks per year or between 10 and 11 trucks per day, plus innumerable small trucks including pickups and small dump bodies), it's impractical to close the doors, and the wind will grab any loose trash near the doors and send it flying through the air to land within a quarter-mile radius. This is where Big Mike came into my life.

Big Mike was so named because, one, he was a big guy, and, two, we already had a Mike and we needed a way to distinguish which Mike we were talking about. Big Mike was hired by the behemoth trash company running the city's transfer station to collect the windblown litter in and around the facility. Seven and a half hours a day, Mike pulled a ninety-six-gallon trash bin on wheels behind him as he made his rounds. He was by no means speedy, but he was methodical. Big Mike, dressed in jeans, t-shirt, boots, and an orange safety vest, would put on his earphones and get lost in music until lunchtime, when spending time with his affable personality became something I looked forward to. Big Mike was in his early twenties. He had started college, but that didn't work out. He was a musician, and he brought his guitar in one day, revealing himself to be a natural bluesman.

Lunch with Big Mike was also fascinating because this young man had quite the appetite. Most of the lunch crowd had little lunch boxes with a sandwich, a piece of fruit, some chips, and a soda. Big Mike brought in a full grocery bag that had several sandwiches, something he had to heat up, a big bag of chips, a half-gallon of soda, and gooey chocolate cake for dessert. Big Mike wasn't obese, but more of a large-framed colossus. I'm guessing he walked six to seven miles each day of work, so he burned a lot of those calories.

Once a week, we would splurge and get takeout from the local burrito joint. We referred to them lovingly as belly bombs. They were delicious and very filling, weighing in at about a pound and a half each. I'm not sure what came over me, but one day the lunch discussion turned to our burrito order and I began to speculate that somehow I could consume more belly bombs than Big Mike. One burrito was typically enough to put me into a full-on food coma, and every time we ordered burritos Big

Mike would get two. Twenty years prior, I once stuffed an entire Big Mac into my mouth on a dare, and another time I drank a frozen margarita in under thirty seconds (my prize being another frozen margarita, which was very difficult to drink because of what felt like an ice pick stuck in my brain, a lovely byproduct of rapidly consuming a frozen beverage). Once talk of a challenge started, I found my nonsensical competitiveness rising and I started smack-talking Big Mike, saying something like "The bigger they come, the harder they fall."

The following Tuesday, the throw-down was on. Nobody bet on me, which in hindsight seemed like a wise choice. Big Mike had at least one hundred pounds on me, but if something tastes good to me I can eat a shocking amount. Just ask my then spouse about the time I ate nineteen tacos al pastor at an outdoor restaurant in Cuernavaca, Mexico. The waiter looking at me incredulously as I placed order after order. Now I was poised with three belly bombs in front of me—four and a half pounds of Mexican food yumminess and I was feeling hungry. The first went down effortlessly, Big Mike keeping pace. I tore into the second one, adding a dollop of jalapeño salsa to each bite for good measure. As I neared the end of the second burrito, I didn't want to give any indication that I was getting more than full and with a look of false confidence at Big Mike I opened the third burrito with a flourish. I choked my way through half of the third burrito and I noticed Big Mike was slowing down. A fourth burrito lay in wait for each of us should the need arise, but this contest would be decided on the third burrito, and I noticed a spark of panic in Big Mike's eyes as he noticed I was ahead. Another six bites and my third burrito would be history, but there just wasn't any room. I could tell Big Mike was going to get through the third one because his fans were encouraging him and I think he felt obligated to put down the insurrection. We both suffered for our efforts. Going back to work was painful, but the results were congruous with the way things ought to be. Every time I eat a burrito (only one), I think of Big Mike and I smile.

What I couldn't know, didn't know, wish I knew, was that Big Mike was having some mental health struggles. He seemed a little lost having dropped out of college, and the trash-picking job wasn't a career track for him, but he was so earnest and personable that I started trying to figure out how to get him on my staff. There was going to be an opening in a few months, and I wanted Big Mike to get his foot in the door and become a full-fledged member of our team. A few weeks after our burrito contest, Big Mike wasn't at work at the start of the work week. Then again the next day a no-show. He didn't actually work for the city, so he wasn't my responsibility or employee, but I was concerned since he was usually very reliable. On day three, we made contact with his family and learned the tragic news that Big Mike had taken his life. We were devastated by this news. We all questioned what we might have done to help him.

Big Mike's family came to visit us after the funeral. We learned that Big Mike considered us to be an extended family. In a ceremony, we placed a statue of Mary on the hillside above the Recycling Center. Mary was particularly iconic to Big Mike, being a devout Catholic. He is loved, he is missed, and he lives in my heart forever.

Chapter 16

The same week I learned of Diane's father's death, I found out that my former boss, the public works director who hired me, Bob (yes another Bob), had also passed. In truth, it was somewhat miraculous that Bob had lived to eighty-five, as he had been a several-pack-a-day smoker long before I met him thirty years ago. Also miraculous is the fact that Bob even hired me in the first place. In spite of being from veritable different planets, Bob and I got along famously, even if we had fundamental disagreements as to how we viewed the world.

Bob was serious old school. Prior to becoming public works director in the late 1980s, Bob had spent decades as an engineer with a huge construction firm and was a project manager for large portions of the interstate highway system that cut a huge swath through the mountains and over the many rivers of West Virginia. Sensitivity about environmental issues was not even an afterthought in those days of highway construction, so his hiring of me, a guy who wears his environmentalism on his sleeve, seemed incongruous.

As noted previously, one of the questions Bob asked me during my interview was if I had twenty tons of green glass and I could pay $40 a ton to have it recycled or bury it for free at the dump what would I do? I'm a fairly pragmatic person in general, and I don't often see things in black and white, but rather in shades of grey. As it turns out, I was fresh off a voluntary internship at the EPA where my work focused on green glass recycling so I had a ready answer for Bob. I explained that I found it difficult to believe that those would be my only two options, but under the scenario he described I would choose to bury the green glass in the dump because green glass is inert, with no chance of polluting the environment, and that the raw material to make glass—sand and silica—are abundant, so it didn't make much sense to me to spend $800 to recycle the glass. The waste of money trumped the waste of resources. Maybe Bob wasn't expecting this answer from the tree hugger like me, but he seemed to like my reasoning. All I knew was that there were fifty-seven other applicants for this job, including several attorneys and a handful of PhDs, so I thought the chances were slim that I would be selected.

Selected I was though, and my first few weeks on the job I was wide-eyed and quiet, trying to absorb the responsibilities of the job. Bob would frequently corral me to join him as he made the rounds about the city. Keene has 152 miles of roadways, and Bob and I drove many of them, officially to inspect the road conditions—a culvert or a catch basin, potholes and signs—but mostly it gave Bob an opportunity to smoke. The city had a no smoking policy, but it wasn't exactly enforced in city vehicles, and because I had not yet found my voice, I suffered in silence, windows up, air conditioning blasting, and the slow, slow burn of Marlboro cigarettes that miraculously dangled from his lower lip as he chatted with me. I'm horribly allergic to cigarette smoke and I have asthma, so I was sure Bob was cutting years off my life as I inhaled his secondhand poison. A couple times, I might have squeaked something about please don't smoke, but my pathetic pleas fell on deaf ears. Somehow, I became the go-to guy for these local road trips. I loved hearing his stories about his days building our amazing interstate highway network. If you've ever seen the scale and scope of a minor highway project and then realize that he oversaw the construction of hundreds of miles of roads and dozens of bridges, this man was in charge of thousands of employees, hundreds of pieces of equipment, and many millions of dollars of materials. He literally moved mountains. Keene must have seemed pretty sleepy to him, but he was on top of it all with his eyes closed and Keene was better for it.

I loathed sitting in that smoke-filled car. Keene became interested in an emerging composting technology that was combining wastewater treatment plant sludge (biosolids is the new term, to make it sound less yucky) with mixed household waste to produce a soil amendment that was purported to have a variety of uses such as landscaping or even agriculture applications. The nearest operating facility was located in Pigeon Forge, Tennessee, which, incidentally, is also the home of Dollywood, the theme park created by country music legend and all-around stellar human, Dolly Parton. Pigeon Forge is at the gateway to the Smokey Mountain National Park which is an absolutely lovely place to visit, but Pigeon Forge? Well, let's just say that Pigeon Forge is a good argument for why a community should have strong zoning regulations. No disrespect meant to the fine folks of Pigeon Forge, who I shamelessly hope will buy this book.

To get to Pigeon Forge meant a road trip of about one thousand miles, and since the city wasn't about to spring for airfare it meant fifteen hours in a car, with Bob, filled with smoke. The plan was to leave first thing Monday, arrive sometime on Tuesday, turn around and be back late Wednesday. Bags were loaded in the trunk and I walked over to the passenger door. Bob was in the car, the ignition was on, and I crossed my arms and waited outside the car. About ninety seconds later I got a reaction. Bob opened the door, stood up, and hollered over the roof "Dunc [he always called me Dunc], get in the car!" My arms still crossed I answered, "No," quite firmly. "What do you mean no?" Bob asked, incredulously. "What I mean is that if you're going to smoke I'm not getting in that car," I said. "Just get in the car," Bob ordered as he climbed back in. I stood my ground and a moment or two later Bob got out and said, "You're really not going to get in the car are you?" "No, I'm not," I repeated. I told Bob we could stop as often as necessary for smoke breaks, but that it was against city policy to smoke in a city vehicle and unless he agreed not to smoke we'd have to make alternate arrangements.

Every thirty minutes or so we pulled over and Bob would get out for a smoke break, his gravelly voice regaling me with the tall tales of his big construction days, and I knew he wasn't exaggerating as we literally drove on some of the roads he had built all those years before. Bob never smoked with me in the car from that day forward, and strangely it seemed to deepen his affection for me. He always treated me like a favored son. The compost technology turned out to be a bust. Too much broken glass was making it through the compost process, and shards of glass were plainly visible throughout the finished product, which they were using as a daily cover at a local landfill. Not exactly the revolutionary process and product we were hoping for. I got fifteen hours of smoke-free stories of his version of pyramid building on the way home. It was a great road trip.

Other road trips were soon in the offing, as some of my projects began to work their way through the political process. Since Bob and I lived in the same town twenty-five minutes north of Keene, Bob decided it would be a good idea if I drove him to the city council meetings in the evening. The work day generally ended at 3:30 or 4, and I had time to scoot home, walk the dogs, get in a workout, have a bite to eat, and arrive at Bob's house by 6 to make it to the 6:30 meetings at city hall. It didn't take long to figure out that Bob was being pragmatic as well because my chauffeuring allowed Bob to fully enjoy his five o'clock cocktail hour without having to worry about driving later. I was very prompt, the trains in my life usually ran on time, and I pulled into Bob's driveway in my bright red boxy Volkswagen Jetta with a roof rack that was used to transport bikes or skis depending on the season. The bars of the rack extended about eight inches beyond the edge of the roof. Getting in my car with this protrusion required a modest amount of care to avoid being struck by the metal bar, but Bob, happy hour in effect, managed to score a direct hit in the middle of his forehead . . . every . . . single . . . time he got in my car. And I was entertained by the same expletives as he wondered why I would have such a hazard attached to my car. Simultaneously, I wondered how he managed to hit the bullseye each and every week. I began to anticipate the hilarity when I pulled into his driveway and he never failed to disappoint. There was many a meeting where Bob addressed the city council with a welt smack dab in the middle of his forehead. I wonder if the city council ever noticed.

I only got to work with Bob for a couple of years. He took a well-earned retirement when he turned sixty-five. I would see him around town a few times a year thereafter and we'd catch up on each other's lives. He seemed proud of what I had accomplished each time I saw him, and I thanked him for taking a chance on me. I'll always be grateful I had an opportunity to work with this great man who might have had a gruff exterior, but I saw that he had a heart of gold.

Chapter 17

In the earliest days of my tenure at the dump, our Human Services Department had a welfare to work program. If you were out of work, needed funds for housing or food, and were able-bodied, then you had to work to receive these benefits. Every day around 8 AM, I would show up at the Human Services Department to pick up that day's crew. Some days there were none, some days it could be five, but for quite a while we ended up with a consistent cast of characters. I brought them up to the Recycling Center and they would augment my regular staff with much needed additional bodies.

There was a definite daycare aspect to supervising the welfare workers, as they were creative in the ways they could conspire to avoid actual physical labor, and the smoke breaks or secret nap spots had to be monitored lest they get out of hand. All in all, they were interesting personalities with more than their share of hard-luck stories. One of our most consistent welfare workers was Warren, who was purported to have a steel plate in his head. Some of the workers tried to lure him close to some of our industrial magnets we used for scrap metal, but I never saw him overtly magnetized. Warren talked a big game and his prized possession was a steel thermos that he brought with him filled with coffee. He loved to brag about how it would keep his coffee hot for eight hours, and was, in his words, absolutely indestructible. Examining his thermos, I had to agree that it was substantial, and I could clearly see the steam rising as he poured his last bits of late afternoon coffee. I never doubted that his thermos was also indestructible. But his fellow workers were determined to test the limits of his thermos and one day when Warren wasn't looking they grabbed it and placed it under the wheel of the loader just before it backed out of the building. Indestructible it was not. By the time the thermos appeared on the other side of the tire it was as flat as it could possibly be. There was quite a commotion as I emerged from my office trailer, seeing the flattened thermos and Warren going ballistic on his co-workers. I decided Warren needed to cool off and he was done for the day. I directed him to get in my car to take him back downtown. Even in his extremely agitated state Warren managed to avoid the bar of the roof rack as he got in my car, but somehow he managed to get himself tangled in the shoulder belt of the car, that was attached to the door in an attempt by Volkswagen to increase seatbelt use. The idea was just to slide in behind the belt, close the door and voila, shoulder belt already in place. Except Warren had somehow managed to wrap the belt around his neck and the door wouldn't close, but every time he opened the door the noose would tighten around his neck causing his eyes to bulge, his tongue to stick out, and a strange high-pitched sound to come out of his nose (probably expelling air). I stood there mesmerized because Warren just could not seem to fathom the cause and effect of his current situation as he repeatedly tried to open the door, more forcefully each time, but with the same result to his eyes, tongue, and nose. Granted it was a little mean to watch Warren do it over and over, but recalling it all these years later still makes my shoulders shake with laughter. Before Warren left for greener pastures I did get him a replacement thermos, but I suggested he tone down the indestructible rhetoric.

Chapter 18

I care. That's the curse. Most of the time I don't say anything in spite of my interest in the sociology of trash disposal, but the bottom line is that I care. It bothers me when people are careless about resources.

Years ago, I was walking down the street with a friend chatting about this and that. As he finished the soda he was drinking, he tossed it in a trash receptacle with as much thought as if he were swatting a mosquito. I'm thinking to myself, okay, that's equivalent to about eight ounces of gasoline in wasted energy. I doubled back to retrieve the can, but didn't say any more about it.

I've attended many a social event where everything—bottles, cans, paper, plastic, organic waste, etc.—all went into a trash compactor. The sound of glass being crushed and cans being flattened into a wasteful mass makes me despondent. I lay out every argument I know (which is a lot) in favor of recycling and or composting over what is being practiced, but usually I don't make much headway. The sad part is that the children are likely to learn the practice of discarding resources from observation. In the end, I usually put all the bottles and cans from a gathering on the counter and I take them home to place in my recycling bin. At least a handful of recyclables are spared an ignoble fate. If there is a large amount of organic material—think watermelon rinds or leftover corn cobs—I have been known to stuff a bag with these compost jewels and bring them home to be put in my compost pile.

It's always shocking to me when I'm away from my carefully controlled disposal system that first looks to recycle or compost as much as possible, leaving refuse as the last resort, to discover that people, even people who should know better, are so cavalier about their discards. Without an extreme effort, I manage to recycle or compost over 80 percent of my home discards. Granted, not everyone has access to a backyard compost bin, which can account for a shocking amount of discard diversion.

Each spring, as my compost pile begins to thaw out, I have visions of harvesting my first batch of the season. This activity, more than any other, gets me into the gardening mode. I love the texture of finished compost and that rich earthy smell. My imagination can easily see the tender plants and seedlings bursting forth as compost is added to the soil. Any type of plant can benefit from compost.

Put simply, composting is a biological process whereby organic material breaks down (decomposes), courtesy of microorganisms and invertebrate animals who consider your discards a delicious delicacy. As they consume the organic material it breaks down into smaller compounds, carbon dioxide, and water. Once the decomposition process is complete you're left with a soil-like product that, when added to soil, enhances the ability of the soil to retain moisture, improves soil structure, and attracts earthworms which in turn aerates the soil which allows plants to grow more vigorously.

Pretty much anything organic can be composted under the right conditions. An extreme example of this was a field trip I took many years ago to a seaport town in Maine where sea urchins were grown in an aquaculture operation and after the meat was harvested the shell of the sea urchin remained. Surely you've been to the beach at low tide when the air is rich with the mixture of salt, fish, and shellfish. A lot of people love that smell, and to me it's very evocative, as my connection to the ocean and beach is very strong. Imagine that low tide smell ramped up about a thousand times and you begin to have a sense of what a sea urchin compost facility smells like. Housed in a negatively pressurized building (air does not escape the building, but is sucked back in, even when a door opens, for purposes of minimizing the migration of odors, and boy is that a good idea in this situation), the carcasses of the sea urchin are brought in by the thousands and placed at the beginning of a row. Some amendment is added, such as wood chips or wood ash, to provide the optimum medium to produce a finished compost. A flail that rides on top of some rails on the concrete walls that form the rows mixes the sea urchins and amendment and slowly pushes the material forward, creating room for the next deposit of sea urchins and pushing the finished compost off the end of the pile after something on the order of a twenty-one day "cooking" process.

I couldn't smell the operation as I walked toward the door, but once inside there was a veritable fog created from the moisture evaporating off the sea urchins. The military or police ought to consider bottling up that smell for use in dispersing crowds, because the smell assaults you, buckles your knees, and it's a struggle to breathe as you can taste the smell as it is forever burned into memory. How a worker could get used to working in that place is simply beyond comprehension. I've been in the waste business for thirty-two years and smelled all kinds of nasty things, but the decomposing sea urchins ranks #1 as the nastiest. Once you smell something like that, you'll never forget it. It makes a helluva compost though—black gold.

Composting Guidelines

Materials to **Include** in Backyard Compost

Aquatic plants	Garden trimmings	Sod
Bread	Grass clippings	Tea leaves
Coffee grounds	Hair clippings	Twigs and shredded branches
Egg shells	Leaves	
Farm animal manure (e.g., sheep, cow, horse, poultry)	Soiled or nonrecyclable paper (shredded)	Vegetables
		Wood ash
	Sawdust	Wood chips
Fruit	Straw	

Materials to **Exclude** in Backyard Compost

Bones	Fish scraps	Peanut butter
Pet manure (e.g., dog, cat)	Lard	Salad/cooking oils
	Mayonnaise	Salad dressing
Dairy Products	Meat scraps	
Diseased plants		

Source for table: "Backyard Composting," Municipal Solid Waste Factbook, US Environmental Protection Agency, no date, accessed October 18, 2023, https://p2infohouse.org/ref/03/02064/factbook/reduce2.htm. Adapted from "Solid Waste Management," the October 1993 Newsletter (vol 7, no.10) of: The University of Illinois at Chicago Office of Solid Waste Management (M/C 922).

Whether in a loose pile, a homemade bin, or a commercially sold compost bin, you can make compost. It's easy, fun and provides a valuable soil amendment for your flowers, vegetables or lawn.

Ideally, compost piles should be at least one cubic yard in size (a pile 3 ft. by 3 ft. by 3 ft.) because that mass provides enough heat retention capability for the microorganisms to thrive. If you don't happen to have the necessary sized container, organic material will still break down over time, it just takes longer. Unless you're a reasonably experienced composter and you understand that adding certain things to your compost can have a negative impact on the compost process such as attracting unwanted pests, or the worst outcome, the compost pile goes anerobic and starts to smell, it's best to stick to the basics.

The following recipe is a rule of thumb, and you can make a lot of errors and still achieve your objective, but using the ratio of three parts brown to one part green, is reasonably foolproof. The "browns" are carbon sources, the most abundant being leaves, and the "greens" are nitrogen sources, typically grass clippings or pre-plated fruit and vegetable scraps.

I know many people who have a compost bin and they do nothing more than throw the browns and greens in and don't give it another thought until a year or two later when they access the finished compost from the lower levels of the compost bin. If you're a little impatient, like me, you can turbocharge the process by mixing the browns and greens once a month, and ensure that the moisture content of the material is akin to a wrung-out sponge.

It's best to avoid putting in any meats, bones, dairy, and oils into a backyard compost pile, primarily because a typical backyard compost pile cannot generate enough heat to break down pathogens, and we'll just leave it as that's bad. A few years ago, I fashioned an in-ground composter that could take meats, bones, dairy, and oils, and everything else typically considered compostable. I purchased a round plastic laundry basket and cut out the bottom of it. I dug a hole in the ground and put the laundry basket in, put a piece of wood over the top of the basket with a hole cut in the wood for access, then placed an old traffic cone on top that was trimmed about four inches below the top to make a larger opening to feed the bin. I backfilled the dirt from the excavated hole so that the only thing visible was the traffic cone and the coffee can that I put on top to cover the opening. I didn't need to do anything other than add organic material to the bin. This won't be the kind of compost you harvest per se, but the organic material WILL break down (there are a lot of very small things living in the ground that will devour most anything you give it). I've never encountered any odors or unwelcome guests wanting to investigate. I was, however, amazed by the results, and I no longer had to contend with stinky trash bags.

You like the idea of composting, but don't want a compost pile in your backyard? Many locales, such as Keene, have giant compost piles primarily consisting of leaves collected in the fall along with a small amount of grass clippings and other organics. Keene allows pre-plate food to be composted in our compost pile. Because our pile is big enough, it generates a fair amount of heat and the material in the pile breaks down to allow for harvesting after about a year without doing much except creating the pile in the first place.

Of course there's another way to manage organic waste. The folks who work at the Recycling Center are a sociable and frugal bunch. We've refurbished many a gas grill over the years, and when people drop off the propane gas cylinders for recycling often there's enough gas left in some of the tanks to fire up the grill for lunch. One such opportunity for a hot dog cookout had the usual spread—hot dogs, buns, condiments, and my favorite: chopped onions. I piled on the onion and after I had consumed a couple of hot dogs, a former supervisor, Charley, announced that he was glad he had spotted the onion coming down the recycling line because "After all, what's a hot dog without onion?" Needless to say he was the only one who was pleased with where he got the onion, but we all seemed to be none the worse for wear as a result.

My message about the fundamental importance of recycling and composting is not meant to be holier than thou. I'm not better than you because I recycle or compost most of my stuff. To me it's an ethic that belongs in the same league as opening doors for people, saying thank you and please, letting someone go in front of you in the checkout line, or brushing your teeth. It's golden rule category. Beside the economic arguments and the environmental benefits, recycling is the right thing to do.

In this complex world we live in with all the environmental issues that seem out of control, recycling is a positive action that you can take that does make a difference. Since 1994, the Keene Recycling Center has processed over 250 million pounds of recycling, and many millions of pounds of compost (we don't weigh the compost). We've sent out over 6,000 tractor-trailers of recyclable material which, lined up end to end, would stretch 62 miles. And just to prove that environmental initiatives and economics are not mutually exclusive concepts, the city of Keene, a hamlet of just under 24,000 people, have generated over $10 million in recycling revenue and avoided disposal costs of over $25 million that would have been incurred if those recyclables had simply been discarded. If it's a curse that makes me care, I can live with it. All I know is that recycling and composting makes more sense than burying it in the ground. I have to laugh over that last sentence because composting is technically burying organic material to decompose, but the point is that burying organic material in a landfill isn't the highest and best use of the resource that composted organic material represents.

Chapter 19

With a population of close to 330 million, the United States is, in fact, the world leader in producing waste per capita. Annual waste generation is somewhere north of 250 million tons. Each individual in the US produces close to 5 pounds of waste each day compared to about half that rate (2.6 pounds per person) for the rest of the globe (these EPA statistics are likely severely short of the actual US generation mark, which new evidence is pointing to a significantly higher per person daily generation of waste). I'm not sure that our profligacy is a requirement of our affluence, but the US likes to be a world leader and we have got this one in the bag. And yes, waste is a waste, however, it's important to remember the fundamentals of the waste business. When someone scrunches their nose when visiting the dump, I remind them that it isn't trash they are smelling, it's money.

In the earliest days of organized trash collection, driven by public health advocates to help reduce the incidence and spreading of disease, trash was a business that was easy to enter and, with a bit of savvy and sometimes a bit of muscle, it could become quite lucrative. Towns or cities that opted to allow the private sector to collect and dispose of waste created a haven for organized crime to create a veil of legitimacy. The business itself is legal, and if there was a public contract attached to the business it assured big profits. If organized crime decided they wanted control of a territory they would buy up the competition, fix pricing, and effectively give their customers no choice to choose another hauler. New York City became the largest city in the US to have the trash business controlled by organized crime. Beginning in the 1950s, members of the crime syndicate infiltrated the Teamsters Union. Trash-truck drivers were members of the Teamsters, and by controlling which companies a driver could work for it became relatively simple to drive out non-organized-crime-affiliated companies.

This stranglehold of organized crime's monopoly in trash began to unravel in the early 1990s in no small measure due to the efforts of Rudy Giuliani, mayor of New York City, who did a credible job wresting control of the collection of trash from the crime families and back into the public sector. (Rudy's good deed is now largely forgotten given his cringe-worthy alliance with the twice impeached, multiple criminally indicted 45th President and his highly questionable "legal" efforts to contest the 2020 election.)

It was during this organized crime to public sector transitional period, in 1992, that I received a call from Vinny. I don't know if you've ever met someone with a true Long Island, NY, accent, but having spent some time on Long Island in my youth, it is immediately recognizable (I also briefly dated a woman from Long Island in college, and while I can't say that we broke up because of her stereotypical nasally accent, it didn't help either). Vinny wanted to discuss some business with his first question being, did I have the key to the landfill? As the manager of the dump, I answered in the affirmative. Then Vinny got right to the point. "Would you be interested in some after-hours deliveries?" Curious, I asked what he meant by after-hours. "Maybe two or three in the morning, probably one load a day (a hundred-yard trailer), sometimes two." I was silent. "And there would be around $1,500 per load for your trouble."

They say everyone has a price, and perhaps that's true. I enjoyed a few moments of fantastical thinking. The amount of money was easy enough to calculate, and so was the amount of time it would take to get caught. Yes, the landfill was reasonably well-hidden by a thicket of trees, but any truck traveling through Keene at 2 AM would arouse suspicion, and not only was the prospect of going to jail entirely unappealing, there was absolutely nothing right about this offer on the table and everything wrong. It's a strange thing to think about being a phone call away from organized crime. It just goes to show that trash is big business and unsavory business relationships are closer than you think.

Several years later, a landfill gas incident brought that brief conversation with Vinny back to mind. After the landfill was closed and capped, we installed solar-ignited gas flares on the oldest sections of the landfill. The areas where the gas flares were located had first received trash forty years earlier and there wasn't a huge amount of gas being generated due to the fact that most of what would decompose had done so. Occasionally gas pressure would build up in these areas and the flare, designed to sense the pressure, would open a valve and ignite the gas. This wasn't very impressive during the day because the flame didn't show up too well, but at night it was fairly dramatic. If you happened to be nearby when the flares would ignite, there would be a clicking sound followed shortly by a whoosh as the combustible gas (methane) ignited. Some nights when I was driving home after a meeting I could see the flares, which looked exactly like the Olympic flame, glowing orange as the light danced through the leaves. Of course I knew what was going on, but an intrepid volunteer fireman who happened to be driving home one night noticed the unmistakable glow of fire and he did what every fireman would do—he responded. Except he didn't know about the chain that had been strung across the road because the landfill was closed. He raced down the access road heading toward the flame like a moth to light and practically tore the roof off his car. He's actually lucky he wasn't decapitated. Undaunted by the damage to his vehicle, he walked the rest of the way into the facility towards the flame only to discover the controlled burning of excess landfill gas.

But had I entertained Vinny's offer and done the illegal, unethical, and immoral deal, it would have been the volunteer fireman, or the curious passerby who happened to have insomnia at the wrong time who would have revealed the truth. There's just no getting away with something like that, and knowing that helps me sleep at night. If everyone does have a price, I'm glad it hasn't been me that's found out either through my failings or the temptations of my colleagues. Vinny, if you're out there, yes, we do take deliveries, but only during business hours, and we don't take cash, we only accept checks.

Chapter 20

There are literally thousands of stories that begin or end at the dump. Many of them are humorous, and, frankly, we need more frivolity in our serious lives. The following are mini stories, not a chapter unto themselves, but perhaps it's a pet peeve, or an issue where something that seems black and white is more like shades of grey. These stories need airing, so here it goes.

First there is the unmistakable smoke, pure white or off-brown. Then comes the odor. Cruising around the Monadnock region on my bicycle on a cool fall day recently, I noticed several homes were actively burning household trash in a burn barrel. People have been doing it for generations, but what most people don't realize is that the constituents of the trash have changed dramatically over the years. Plastic products now represent approximately 12 percent (and growing) of the typical waste stream and burning those and other items in the burn barrel creates a toxic stew.

A study conducted by the American Chemical Society found that a family of four burning trash in a barrel in their backyard can potentially put as much dioxin and furan into the air as a well-controlled municipal waste incinerator serving tens of thousands of households. These polychlorinated compounds can be formed simply by burning common household trash at low temperatures.

Your trash seems innocuous enough. The composition of household trash burned in a burn barrel usually includes newspapers, magazines, junk mail, cardboard, milk cartons, and various types of plastic. The results of the American Chemical Society study did not include the emissions of burning paint, grease, oils, tires, or other household hazardous wastes, but often these are thrown in as well.

I know several people who make one trip to the recycling center per year to dispose of accumulated recycling and refuse. The collection is impressive; the pickup truck is almost always piled high with glass bottles, steel cans, aluminum cans, and a few bags of refuse. I'm curious about this, so I ask how they do it and the response is always the same. What could burn was burned, those items that don't burn or compost are either recycled or thrown out as refuse.

It's easy to think that your small amount of burning can't possibly hurt anything. The American Chemical Society study indicates otherwise. The average family of four produces almost 6,000 pounds of waste per year. If burning the majority of this waste produces emissions equal to the emissions from a modern waste incinerator with millions of dollars of pollution control equipment, what is the cumulative effect of dozens of people burning household trash? Can't be good.

Although the adverse effects of exposure to certain dioxins in people is not fully understood, it is believed that immune dysfunction, cancer, hormonal changes, and developmental abnormalities could be related to dioxin exposure. Common sense would say that exposure to the smoke from your burn barrel is bad for your health. Chances are, however, that if you burn trash in your backyard, you are standing upwind of the smoke. But what about your neighbor who lives downwind?

And then there's littering. There's two types of littering: inadvertent and deliberate. The main route to the dump has always had a litter problem because many people either didn't take physics in high school or weren't paying attention. From March through November, there is a crew of people who patrol the main travel corridor, about four miles worth, on a monthly basis to collect the litter that blows out, mostly, from the pickup trucks that travel to the dump. Part of me "gets" it. When you're getting rid of something, it isn't exactly a high priority to take care of the item you no longer want. And this is where the physics comes in. If one haphazardly places refuse loose or in bags in the back of a pickup truck and subjects it to the wind currents generated at fifty–sixty miles per hour, things are going to fly. This is the inadvertent type of litterer. It wasn't the intent, but the outcome is the same, and it keeps a lot of people very busy a day each month, costing hundreds of dollars each time they are deployed to pick up the stuff that blows out the back of the pickup trucks. It's annoying that month after month it's the same drill. We even remind the folks who arrive at the dump with an improperly covered load that local laws require covers to prevent trash from being windblown, but it's déjà vu all over again each and every day.

The deliberate folks are a little harder to figure out. I love to bike, and being in a rural area, the scenery includes the occasional backyard burning operation as mentioned above, and sometimes in very isolated areas I discover the troves of small dumps where all types of items can be found. Household trash isn't common; it's usually items like mattresses, shingles, tires, or paint cans that I stumble upon. Occasionally, I do find bags of household trash, which I can never help examining because I hope to find identifying information with which to help catch to the scofflaw. The story of identifiable household trash being dumped off the beaten path often has a common theme—a guy (yes, it's always guys) in a pickup truck is hired to haul some stuff away, and instead of having to pay a small amount to get rid of it at the official dump, he finds a favorite out-of-the-way place on a backroad and those six bags of stuff I find represents twelve dollars he didn't have to pay. Now there are a lot of people, like me, who find illegal dumping to be morally reprehensible. It's most certainly a pet peeve of mine, and I do take delight in turning over evidence of illegal dumping to the authorities, as almost every town has fines for littering. I don't know what percentage of people get caught, but if bags are being chucked, the chances go way up versus the unidentifiable mattress, shingles, or tires. One particular unknown person takes the prize for determination, though.

I came across several bags of trash along a fairly well-traveled road one day and opened them up to find that every single shred of identifying information had been meticulously covered over with an indelible marker. I'm talking hundreds of pieces of paper and other identifiers and each one gone over multiple times with a marker. It must have taken that person many hours to remain anonymous. He saved about eight dollars, though, so it must have been important. The runner-up was the person who ripped up anything with an address into tiny pieces, so tracing was impossible. I'm sure these are not bad people, but they are pretty misguided.

And speaking of pet peeves, let me just add cigarette butts to the mix. The fine for littering is somewhere on the order of one hundred dollars, and we could have solved our municipal budget issues many times over if our littering laws were enforced on the back of cigarette butts alone. Spend a day noticing how many people simply toss a cigarette butt on the ground or fling it out the window and the awareness will astound you. I have a colleague who was actually fined for littering for tossing a cigarette butt out the car window, but he's the only person I know who was ever fined. It's literally impossible to clean up all the cigarette butts that are carelessly tossed into our environment, and many, many of them make their way into our waterways—streams, lakes, and oceans. I think people are under the misconception that the cigarette butts will decompose, but they really don't. Cigarette butts aren't the only thing showing up on the streets. Notice the pattern some time—travel along almost any road in the community you live in and observe the fast-food containers and the cans and bottles that didn't just blow out the back of a pickup truck. A window was rolled down and the stuff discarded without any thought to the action. It is my opinion that this is learned behavior and no amount of education or awareness is likely to change this. For those who exhibit this behavior, their action is equivalent to swatting a mosquito. Here is something for which I have no use for—swat, or in this case, fling. And those with a pet peeve for this sort of thing or those with a paying job to deal with it (or both), try to pick it up one piece at a time.

I'm not sure what to make of this statement, but recycling is more popular than democracy. On a daily basis, more people participate in recycling than vote. I'll choose to view this by saying that recycling has been integrated into our culture. We don't think about it much. For most, it's simply a chore among many, but it is part of the routine. There's still a long way to go. Some time ago, I did a waste audit in Keene to see how much recyclable material was making its way into the dump. Do you think you could hug 142,000 trees? That's how many trees you could hug if we recycled all the recyclable paper in Keene, because that's the number of trees that must be harvested to replace the paper that is thrown away. We have a long way to go to further recycling and a long way to go to tackle littering. There are lots of good ideas out there to increase recycling and reduce litter. The simple fact is that there is no silver bullet for any of these complex issues. The one thing that remains true is that everything has a story. Some are just better than others.

In 1997, as my then wife and I counted down the days until the birth of our first child, I faced an issue that in some ways sparked the entire debate of America's solid waste disposal "crisis": diapers. Being a practical person, with the birth looming, I put aside my feelings of being strapped to a train hurtling out of control down a track with a big curve ahead, not really sure if I'm going to make the corner, and debated the issue of diapers.

Disposable or cloth diapers, paper or plastic, leaded or unleaded, the world of diapers is far more complex than I had imagined. I could certainly see the practicality and convenience of disposable diapers. I had no less than a dozen of my super environmentally conscious friends explain to me in gruesome detail their noble attempts to steer clear of disposables only to admit defeat for a variety of reasons. I've always prided myself on not only talking the talk, but walking the walk. As with many other dilemmas, a balance must be reached.

Aside from the unsolicited advice from family and friends, in this age of electronics I turned to the Internet for some information about diapers (I would also note that when I first typed "diapers" into a search engine, I learned that there are a number of websites that cater to those with diaper fetishes, but, curiosity aside, I let that be). My sense was to use cloth diapers at home and disposable diapers outside of home. My concern about disposable diapers is the chemical makeup of the absorbent material which makes a diaper a diaper. Many disposable diapers use a super-absorbent chemical called sodium polyacrylate which absorbs and holds fluids in the diaper. This chemical has been linked (according to various online sources, credibility unknown) to toxic shock syndrome, can cause allergic reactions, and is apparently lethal to cats if inhaled (I don't know why a cat would be inhaling this chemical, but we didn't have cats, making this less of an issue).

Since the public generally insists on buying white things, dioxin, which is a by-product of bleaching paper, has been found in trace amounts in disposable diapers. Small amounts of dioxin have been linked to liver disease, immune system suppression, and genetic damage in lab animals. Then there's the issue of fragrances that can cause headaches, dizziness, and rashes. At the time of my research, according to the *Journal of Pediatrics*, 54 percent of one-month-old babies using disposable diapers had rashes, and 16 percent had severe rashes. Apparently, widespread diaper rash is a fairly new phenomenon that surfaced along with disposable diapers. Reasons for more rashes include allergies to chemicals, lack of air, higher temperatures because plastic retains body heat, and less frequent changing because diapers feel dry when wet. A diaper service was beginning to sound better and better. As for using disposable diapers, I wanted to try to find ones that don't contain harmful chemicals, dyes, fragrances, and anything else that might harm our baby.

The fact that the six thousand hospitals in this country generate more than ninety-five million pounds of plastic waste annually was about the furthest thing from my mind as my then wife and I checked into the hospital to finally begin the birthing process eighteen days after the "due date." As my then wife began the labor process, an army of health professionals came in and out of the room, but not before opening and shortly thereafter disposing of plastic objects wrapped in plastic bags that were contained in plastic trays. Mind you, I came down with a vicious flu, which was peaking at 102° as her labor began, so I wasn't in great shape, but even in a state of delirium combined with the emotion and excitement of birth I had two general observations: that there were an awful lot of tubes sticking into my soon to be daughter's mother, and that the amount of waste generated in a hospital is staggering.

Eighteen hours later, we were staring at the most beautiful girl I have ever seen, Samantha Ann. At the instant of her birth, my life changed forever. I have always prided myself with my ability to divert waste through composting, reuse and recycling, but diapers accumulate quite rapidly and admittedly we made a conscious decision to use disposable diapers. Now instead of diverting 80–90 percent of our discards, we fell into the low 70s. Of course, if you are determined, you can rationalize anything. I know that disposable diapers represent less than 1 percent of the overall waste stream, therefore a much more pressing problem, in my mind, was how to decrease the most prevalent item of our daily discards: paper. In our home, we recycle all recyclable paper, compost unrecyclable paper, and generally avoid paper products when good alternatives exist (handkerchiefs versus tissues, cloth napkins versus paper, etc.)

Our second daughter, Selma Marie, arrived in the world late at night in February 2000 with her lungs functioning at full volume (and now my eyes had the two most beautiful girls in the world filling my vision). Many things flashed through my mind as I held this precious being, but one thing I didn't even think of was diapers. That I would use disposable diapers was never a debate. From our first experience, I knew just what we would do. Since my world again revolved around diapers, it only made sense that I take a fresh look at disposable versus cloth diapers.

Disposable diapers cost about $70–$100 per month. Cloth diapers cost about $40 per dozen plus around $70 for a dozen covers. With an adequate supply of diapers, covers, and the cost of laundering it works out to around $40 per month. Unfortunately, the economic debate doesn't factor in some externalities that made disposable diapers the preferred choice in my household.

The solid waste aspect of disposable diapers is not a trivial issue when viewed on a grand scale. It is estimated that upwards of 18 billion, that's billion with a *b*, diapers are disposed in the United States each year. That equates to approximately 3.5 million tons. To put that in perspective, the city of Keene landfill has just under 1 million tons buried in it from several decades of use. If around 1 percent of what is buried in Keene's landfill is diapers, that equates to approximately 10,000 tons or 20 million pounds. Nationwide, diapers represent around 1 percent of the waste stream. One percent doesn't sound like much until you begin to think about the billions of pounds 1 percent represents.

In the case of disposables, it became a matter of practicality. We both worked. Daycare centers will not allow cloth diapers and when we got home in the evening it was about all we could do to feed and bathe the children, take care of the dogs, and do normal household chores. Dealing with cloth diapers, by mutual agreement, would have put us over the edge.

After months of potty training, our family was finally free from diapers. My youngest daughter proclaimed herself a "big girl" and diapers became history. Being done with diapers is like finding money on the ground. No longer did I shell out $70 per month to buy something that contributed so significantly to our solid waste stream.

People have strong feelings about diapers one way or another, but the environmental debate over disposable versus cloth diapers makes the decision to use one or the other even more difficult, because there really is no clear preference. Each option has some major drawbacks. The most significant drawback of disposable diapers is the mere fact that they are disposable. Cloth diapers, for most people, are not as convenient. They are also generally made from non-organic cotton, which is among the most highly chemical laden crops grown in the world. Then there is the energy required to heat water to a temperature that will kill off the pathogens in soiled diapers and the subsequent water treatment requirements.

We have taught our daughters many things throughout their lives. Primary among them was an ethic of resource conservation. Reducing, reusing, recycling, and composting are simply habits. To do otherwise will seem strange to them. Our daughters have taught me many things as well. Primarily, I learned to be more patient and focus on the moment. Twenty-six years later, I still look into their eyes and see the future. What kind of legacy will I leave them? That answer will depend on my actions in each moment.

The few remaining diapers we had left over became props for the girls' dolls. The dolls have long been packed away, probably to emerge when/if I become a grandfather; I don't hold much hope that the dolls will get potty trained anytime soon.

Years ago, I took over the laundry patrol at my house. Whenever I'm in the laundry room, I quickly sort the whites from the colors, load the washing machine, add detergent, and off we go. Obviously, innumerable people are doing the same thing each day, because there is an endless stream of detergent containers that show up at the recycling center from both curbside and drop-off collection.

I purchase detergent in bulk in order to save as much money as possible. A two-hundred-fluid-ounce container yields about sixty-five wash loads. Inevitably, there is a little detergent left in the bottom of the container. I put a little water in the container, swish it around, and then tip it upside down in the washing machine for several minutes. When I come back, the container is completely empty. Unfortunately, the same cannot be said for thousands of detergent containers processed at the Keene Recycling Center, and this has become a problem.

The sorting process at the Keene Recycling Center involves hand sorting plastics, glass, steel, and aluminum cans. As the bin with the detergent containers grows (detergent containers are made of #2 plastic, which is high-density polyethylene), so does a puddle of detergent. We estimate that there is an average of one ounce of detergent left in each container. Over the course of a year, we process approximately 280,000 pounds of detergent containers. Each container weighs about half a pound, which totals 560,000 containers. If each container has one ounce of detergent and a load of laundry requires about three ounces of detergent, the remaining detergent represents over 186,000 loads of laundry that could be done with the detergent remaining in containers brought to the Recycling Center. At $0.015 per ounce, this represents a little over $8,000 in cost.

The numbers themselves are startling, but the real problem comes when we sell our material to the factories that are using our raw material to make new material, running the risk of having their equipment literally gummed up from the remaining detergent in the container, and this threatens the ability to recycle this commodity. This is, however, one of the situations where the introduction of the detergent pod and their increasing use is changing the face of this quandary. No more detergent bottles with residual soap. Pods in a bag with only the empty bag (which will almost certainly be marketed as recyclable, but in reality has no practical market to effectively recycle) of which to dispose. Will detergent bottles become dinosaurs? Too soon to tell, but we are seeing a noticeable decrease in the number of detergent containers we process.

If you do purchase detergent in an HDPE container, please do your part by rinsing these containers and pouring the residual detergent into the washing machine prior to recycling, and tell your friends and neighbors. You'll save a little money by having the detergent do its job versus leaking in the recycling bin.

Sometimes the stories take a slight detour, as my role with the city of Keene changed some over the years. I have been, for the past thirty years, the Solid Waste Manager for the city, and over twenty years ago I was promoted to assistant public works director, which got me involved in overseeing the Highway Division and Fleet Division. There is likely a *Tales of Public Works* to be written at some point, but there are good stories, and there are great stories. There are three such great stories that I didn't feel should be left out because, well, they're funny.

One evening at a city council subcommittee meeting, the subject of roadkill came up. Of course, we've all seen dead animals on the side of the road, and it is either the city's responsibility to collect and dispose of the dead animals or, if along a state route, the state highway crews handle it. I always feel like I have to apologize to the spirit of the animal lying dead on the roadside, as it's a magic fate for the critters that are simply trying to cross the road and are the victims of bad timing. During the discussion of how dead animals are handled, one of the city councilors asked if the collected animals couldn't somehow be skinned, the hides tanned and sewn together to make clothing for the homeless. It would be very bad form to burst out laughing in a meeting where the tone was serious, and it was obvious that this was an earnestly serious question. I bit my tongue and told the councilor that I would do some research and come back with an answer. I learned very early in my career that you don't ever make up an answer if you don't know because memories of an incorrect answer can live forever.

The Highway Division in the city of Keene is full of hunters, and I sought one of them out who I know does hide tanning. When I asked him what would be involved in salvaging the hides of roadkill, he looked at me as if I was insane, for which I can't blame him. After he stopped laughing, he patiently explained the tanning process, which, to sum it up concisely, is an extremely laborious process. I was glad when he then asked me what seemed to be an obvious question, but I hadn't spoken it out loud: What size animal was the cutoff point to consider skinning and tanning? The vast majority of roadkill animals are squirrels. It was with a fit of the giggles that I imagined rows of curing squirrel hides waiting to be tanned and turned into a caveman coat for some unfortunate homeless person. In no way am I making light of the homeless population and their need for warm clothing, particularly in a typical New Hampshire winter. I am simply certain that there had to be a better way to provide warm clothing to the homeless than getting involved in salvaging hides of any sized animal collected from roadkill.

I went back to the council subcommittee to report my findings and it quietly and thankfully slipped into obscurity, but it remains one of the more appropriately spirited, although misguided, things I've ever had to research. I'm only surprised that the highway guys never sewed me up a squirrel cap to mock my assignment, but maybe that's something to look forward to upon my retirement soiree.

The second story involved a road project the Public Works Department was overseeing along one of our major thoroughfares a couple years ago, and there had been several complaints about traffic tie-ups which one of the city councilors wanted to investigate personally. Upon arriving at the site, she saw one of the contractors along the road and pulled over to question him about the traffic tie-ups. This particular person didn't speak English very well, and the conversation ended rather abruptly with the councilor storming off in her car. A short while later, we got a call from the councilor upset because she had been spoken to inappropriately by the contractor and wanted us to investigate why the person, when asked by the councilor who he worked for, told her he worked for "his penis." One of our engineering staff was dispatched to get to the bottom of this, and within a few moments returned to the office practically unable to speak he was laughing so hard. He had found the person the councilor had spoken to and again, in broken English, explained that he understood the councilor wanted to know who he worked for and that he had told her who he worked for. The engineer said, "The Councilor said that when she asked who you worked for you told her that you worked for your penis." The man said, "No, no, I no say I work for my penis, I say I work for 'Martinez'!" Turns out the contractor on the job was named Martinez, and what we had here was a failure to communicate. Every once in a while, a Public Works employee that knows the story spontaneously blurts out, "I work for Martinez."

The third story is about the limits of separating work life from personal life. Early in my career it became obvious that work will consume whatever time you give it, particularly in supposed "off" hours. In my first years I carried a pager and the sense of dread when it went off was never equivalent to the issue for which I was being paged. As my personal cell phone replaced the pager, I relentlessly drilled my staff that if they contacted me after hours, it should only be in the event that someone was actively bleeding or there was a fire. Otherwise, whatever it was could be dealt with the next business day. Several years ago, I took a week vacation and when I returned on the Sunday before the work week, I reached out to two of my direct reports to ask if there was anything notable for me to anticipate when I got to the office. Both of them responded that all was quiet, and I blithely arrived at work when I started getting peppered with questions about the fire at the transfer station. Fire? What fire? Turns out some incompatible material had been delivered to the transfer station while I was on vacation and spontaneously combusted resulting in hundreds of thousands of dollars of damage to the facility. By good grace no one was injured, but this was the third fire we've had at our transfer station, almost certainly started by the mixing of materials that should not be mixed. When I saw my two direct reports, I told them that *that* was one of those occasions that they should have called me to give me a heads up. They both argued that they had handled the situation professionally and they didn't feel the need to bother me on vacation. Fair enough, but guys, if it happens again, please call me so at least I know what I'm walking into the next workday.

Chapter 21

Of course the old saying is, "The road to hell is paved with good intentions," but this story convinced me that the road to hell is paved, good intentions or not. The year was 2011. Climate change had really started to enter the daily news cycle as climate scientists increasingly began to sound the alarm of doom if carbon emissions were not brought under some control. You've heard all the arguments: weather is not climate, climate change is a naturally occurring process, dinosaurs did okay with carbon levels many multiples of where they are today, etc., etc. I think there's little denying, at least from a scientific consensus basis, that we are in a climate change era due to human inputs. Droughts, wildfires, changing ocean currents, increased storm severity, sea level rise, and a number of other troubling trends are forming a pattern that I've heard has two potential outcomes for humans: terrible (best case) and catastrophic. Through the magnificent show *Cosmos*, I learned that if your outstretched arms represented all of time from the Big Bang to present, that shaving off the merest fraction of your middle finger fingernail would represent all of human history. At that scale we are damned insignificant, but within the fingernail shred, for about the last 120 years or so—once the industrial revolution started in earnest—we've had a dramatic impact on our ecosystem, and we show no real sign of the political and economic will necessary to avoid a likely catastrophe. But that's not going to stop some of us from trying.

Some random workday in 2010, I was reading a news clip about a company that was looking to use heat and power (via a landfill gas generator), from closed landfills to power and heat greenhouses that would grow produce for the local region. That kind of idea is my porn. I could already see the project before I even spoke to the developer. After establishing a mutual interest, the city applied for and received a five-hundred-thousand-dollar Climate Showcase Communities Grant from the Environmental Protection Agency which was meant to leverage the grant funds into a carbon-reduction project. The city of Keene, in partnership with Carbon Harvest Energy, proposed to transform the closed Keene Landfill into a demonstration-level project for greenhouse gas management through a combined heat and renewable power plant (CHP) using methane from the capped landfill. The CHP would heat a greenhouse and aquaculture project, produce locally grown vegetables and fish, and produce algae for feedstock and biofuel production. The combination of these technologies would create a more efficient and profitable system than the city's existing methane to energy plant, as it effectively utilizes all available resources. The whole purpose of the Climate Showcase Communities program was to enable other communities to replicate a successful development. I'm a big fan of public/private partnerships where the public sector does those things that it does well and the private sector does the same and the benefit is the intersection of the public/private sector interests that usually are not feasible on their own. I know this will likely cause many of my Libertarian friends to break out in hives, but I have seen how this type of partnership can develop unique and worthwhile projects that have public benefit as well as economic success.

Carbon Harvest proposed to build a twenty-thousand-square-foot greenhouse with aquaculture and plant production for year-round, high-quality fish protein and fresh vegetables, which would be sold locally. Waste heat from power generation would provide low-cost heat to the greenhouse, and the aquaculture systems would provide high value organic nutrients for the plants and the algae. High winter heat demands of the greenhouse are balanced by high summer heat demands for algae processing, so all the heat value would be fully utilized. Combustion exhaust from the combined heat and power plant and waste from the aquaculture facility are recovered to become the nutrient for algae production. The algae systems would be designed to capture all remaining waste and carbon and convert it to algae for production of biodiesel and livestock feed for fish and poultry.

During the early unveiling of the project, the intended acronym KEAP (Keene Energy & Agriculture Project) morphed into the "Fish and Lettuce Project," which didn't demean the intention enough for me to object to this new project name. The closed Keene landfill was an ideal site to build the first commercial-scale integrated algae feed and biodiesel facility in the state of New Hampshire. The project proposed to remove twenty-two thousand tons of carbon dioxide per year without fossil fuel input, with zero waste output, and to convert this source of greenhouse gas into high-quality local food, biofuel, and aquaculture feed. Even better for our grant prospects was the fact that this project was effectively "shovel ready." We identified land adjacent to the landfill that was of sufficient size to site a greenhouse, and within months of submitting our Climate Showcase Communities Grant, we were off and running with a half-a-million-dollar award.

I tend to get ahead of myself sometimes, much like Charlie Brown did in "Be My Valentine Charlie Brown." Charlie Brown was so excited by Valentine's Day that he brought not one, but two briefcases to school in anticipation of the reams of valentines he would receive that day, and with the Fish and Lettuce Project, I had visions of basil, lettuce, tomatoes, and tilapia dancing in my head. This was to be my career capstone project and years from now in my dotage I would revel about how I had this futuristic vision that I conjured out of a sunburst of creative energy, and of course we all knew it would be successful. It even had the lure of up to twenty-five jobs with entry-level positions being significantly higher than the prevailing wage because the economics were so very good.

I will spare you the gruesome details except to say that in my long career I've had a few projects that I naively thought were so good from every angle that no one could possibly be opposed to them, which should have been my first clue that roadblocks lay ahead. As so often happens with opposition, their narrative was full of falsehoods and slander, but it came so fast and furious that we had to fight a rear-guard action and getting the remaining puzzle pieces to fit together in the face of this negative onslaught served to delay, delay, delay. Engineering and design work continued until the idea began to take shape, but the fragile coalition of funding/investors in the project threatened to break apart at any moment. As a show of good faith in the project, the city had a six-acre parcel of land adjacent to the closed landfill cleared of trees ready for earthwork to shape the site per the design. The land we cleared was a mature forest, and it was bittersweet knowing this project would produce significant environmental and human benefits at the cost of six-acres of habitat destruction. The machinery used to clear the site was something out of the Dr. Seuss book *The Lorax*. The "super axe hacker" was a device invented by the Once-ler to cut down the Truffula trees four times as fast. The sun hadn't even begun to set on the day the project started when the last of the trees on the site were sent unceremoniously through a chipper that didn't even hesitate when a forty-eight-inch-diameter oak tree was placed in its grasp. Watching this efficient destruction made my mind wander from Dr. Seuss to the movie *Fargo*, with its famous wood chipper scene. With a wood chipper such as the one grinding through the trees at a blistering pace, *Fargo* would have been a very different movie.

A few weeks later, with nothing to show for ourselves other than six acres of stumps, the end of the Fish & Lettuce project was not much more than a whimper. The investor group Carbon Harvest had assembled lost faith in the project, and a formal withdrawal was on its heels. I was forced to watch the project opponents' dance on its grave with smug "I told you so's." Like so many things we see in our lives, the truth is that it is way easier to blow something up than it is to build it. After a period of mourning, I washed off the failure of Fish & Lettuce and started to cast about for a way to repurpose the EPA grant.

When the idea for Fish & Lettuce was first conceived, we were well into our second decade of capturing and burning methane from the decomposing trash in the closed landfill. The big problem with a landfill gas system is that there is no gauge to tell you how much gas remains in the "tank." Sure, you can make some estimates of gas production rates based on industry data, but when a generator that burns methane begins to hiccup, it's primarily due to the fact that the amount of methane in the extracted gas falls below the percentage necessary to ignite the gas with a spark. Methane that falls below the explosive limit makes it less a fuel source and more of a giant headache. Over the course of the last year of the attempted Fish & Lettuce development, the landfill gas-to-energy system gave us very clear signs that the end of using landfill gas to generate power was near. Ordinarily, in most other landfills with gas-to-energy programs, you simply disconnect the landfill gas generator and plug in to the utility grid. Our problem was that the energy grid with the necessary 3 phase power was more than a mile away, and without 3 phase power, I could not run the recycling center and its machines. Now I had no Fish & Lettuce project and my coveted landfill gas-to-energy program was nearing the end of its life. And this is when I came up with an idea to use the EPA grant and continue the work of resource recovery.

Generators, in general, are configured to run on a certain type of fuel. Everyone has seen either gas or diesel generators. These are common things. Even natural gas generators are reasonably common. Landfill gas generators are specialized, and not suitable for conversion to another type of fuel. But a generator that ran on 100 percent plant-based fuel? This product simply didn't exist. Why? I was less concerned with why than why not? In theory, bio-fuel or bio diesel can replace petroleum diesel on a one-to-one basis. You yourself might know the odd hobbyist who converted an old Volkswagen diesel car to run on used French-fry oil. We had been using 20 percent biodiesel in our fleet for a number of years to rave reviews, so why not install a generator that runs on 100 percent post-consumer vegetable oil to produce 3 phase electricity to run the motors that are used in our recycling center? Why not, indeed.

There was no technical reason why this idea wouldn't work, so I filled out many EPA grant change forms, received a crate of questions back from the EPA, researched and answered all the questions, and within a matter of months had successfully repurposed the Fish & Lettuce grant to one where the city would purchase a +250-kW generator to serve as our primary method to generate electricity. The city contracted with a local electrical engineering firm to help us develop the technical specifications for what should have been an off-the-shelf diesel generator that would use biodiesel versus petroleum diesel. If I only knew then what I know now. I checked my idea many, many times with people with vast knowledge of internal combustion engines, including the city's fleet manager, who assured me that the biofuel would not only work, but would result in a much cleaner emissions profile compared with petroleum diesel, in addition to sourcing what amounts to a renewable resource fuel. The plans and specifications were assembled. The specifications were very clear that we intended to run our generator on 100 percent post-consumer vegetable oil.

At the bid deadline we received one bid. A local generator firm specified a Volvo generator, and the details on installing the generator and the fuel tank were fairly unremarkable at first glance. A purchase order was issued to begin the generator fabrication, and site work to accommodate the new generator and fuel tank began in earnest. Twelve thousand gallons of B100 biofuel was delivered in April 2018, and in May a fault light on the generator indicated a problem with the DEF system. DEF (diesel emission fluid) is a solution of urea and water that is injected into the exhaust stream of diesel vehicles to turn nitrogen oxide (NOx) gases into nitrogen and water. If you're thinking you've never heard much about nitrogen oxide, it's ubiquitously known as "smog." The one and only parameter of an engine burning 100 percent biofuel that is less favorable than an engine burning petroleum diesel fuel is NOx. To reduce NOx, the engine has a device called a selective catalytic reduction (SCR), which injects the water and urea into the exhaust to enable a diesel engine to meet new EPA Tier 4 emissions standards. When the technician arrived to investigate the fault light, he took apart the SCR to discover the emissions pipe was clogged with crystalized and hardened DEF.

A meeting of all the players involved with this project—the general contractor, the engine/generator sub-contractor, the design engineer, and city staff—was quickly assembled, and in short order the finger was pointed at the biofuel being the culprit. It was at this moment that we then found out that the generator that we had purchased was not warrantied to run on anything more than 7 percent biofuel. The engine/generator salesperson, who assured us the engine was designed to run on 100 percent biofuel, was nowhere to be seen at this meeting and we then learned that he no longer worked for the engine/generator company. Shit!

Here I am, $750,000 into a climate demonstration project, and I'm being told that the source of the problem is the very thing I was proposing as a solution. It then seemed perfectly reasonable to test this theory by cleaning out the emission pipe, reinstalling it, set up a tank with diesel fuel, run it through the engine/generator, and see what happens. I've never been so happy to learn that shortly after the test with the diesel fuel was started that a fault light came on. Same drill—disassemble the exhaust pipe and voila, crystalized and hardened DEF. This proved that it was something other than the fuel. I'll spare you the arduous technical troubleshooting that ensued, but eventually the technician sent by Volvo to help diagnose the problem determined that the injection of DEF was at issue and reworked the exhaust system to dose the proper amount of DEF and in subsequent testing the problem appeared to be solved. Little did I know that my problems were about to get much bigger.

There are no regulations that cover biofuel specifically. If you use biofuel, you are regulated by the petroleum-diesel regulations. When normal people buy a commercial engine/generator, they are already in compliance with federal and state air regulations because the engine/generator manufacturer designs, builds, and tests the engines to ensure they meet regulatory standards. When you buy a petroleum diesel engine/generator and then run biofuel in it, you not only void the manufacturer's warranty, but the regulatory folks want to know that what's coming out of the stack is in compliance with regulatory standards. From a regulatory standpoint I get it, but testing the emissions of an engine/generator in the field is not easy, particularly when no testing protocols even existed for the fuel that we were using. For months, communications were flying from the EPA to the state, from the state to the city, and on and on until an agreed-upon testing protocol could be established.

In the fall of 2020, the stack test was complete and under the initial testing parameters, the engine failed to meet the standards for NOx. Crap. Many, many hours were spent analyzing the results, and the failure appeared to occur when the engine was idling. We also discovered that the idling condition occurred only upon startup of the engine and lasted no more than a few seconds before demand was placed on the generator to supply electricity to the adjacent recycling center, so the testing that had been completed did not represent the emissions 99.99 percent of the time. The city proposed re-testing (and you'd fall out of your chair if you knew how much it costs to do field emissions stack tests), the regulatory folks agreed with the new testing protocols, and bingo, regulatory compliance.

After all of the technical difficulties and regulatory hurdles, the engine was put into full time service at the end of 2020, and to this day continues to supply prime power to the recycling center. Just as I was putting this project in my rear-view mirror and making it part of everyday operations, the dreaded letter arrived from the state.

During the multi-year transition from landfill gas-to-energy to design and installation of a 100 percent biofuel generator, until those prime power sources were available, we still needed power to run the recycling center. This is why we had a backup generator. Redundancy. Prime power not available? Simple, go turn on the backup generator, and the recycling center equipment can function. The backup generator was petroleum-diesel powered, and to be used during "emergencies." We can parse what an emergency represents, but in this case, if my prime power generator was not available for whatever reason, and the recycling center cannot operate without power, that, to me, represented an operational emergency. Material never stops coming to the recycling center, and even a single day without power and an ability to process the recyclable material results in a backlog of material that soon makes the stockpile literally spill out the doors. The only alternative in this case is to take the recyclable material to the transfer station and send it out with the refuse. As I explained early on in this book, I did not consider that to be a viable alternative. Without too much thought, I rationalized that any emissions from a generator operating on petroleum-diesel powering equipment to process tens of tons of material each and every day was a far lesser problem than the mass balance equation of the negative environmental consequences from trashing all those recyclables.

I do understand, after three-plus decades of working in a regulated environment, that regulators are all about checking boxes. But when the letter came from the regulators saying that by using our backup generator the city had violated our air permit and they would be turning our case over to the attorney general's office to determine civil fines and penalties, I was deflated.

Fortunately, we were able to calculate that while the number of hours the backup generator ran did exceed our permit, in spite of the state not agreeing that the use of the generator constituted an emergency, the actual emissions from the use of the backup generator did not exceed our permitted emissions limits. As of this writing, I don't know what civil fines and or penalties the city will be assessed, but pushing the beneficial environmental envelope comes at a cost that too often makes me realize that the remedy is worse than the cure.

While this project, funded in part by an EPA grant (how's that for irony?), was meant to demonstrate the viability of fueling a prime power generator with non-petroleum diesel, we remain the only location in the US operating a biofuel-fueled prime power generator. And for the sake of accuracy, and to point out another absurd rabbit trail we encountered in this project, we technically burn B99.9. 100 percent biofuel, which means 100 percent vegetable oil, and it goes by the acronym B100. It costs significantly more to purchase B100 because of something known as a RINS credit. A RIN is effectively equivalency measurement the EPA uses whereby one RIN credit is equal to one gallon of ethanol fuel.

When a company blends fuel it receives a unique identifier number that allows the fuel blender to generate RINs credits that can be traded as a commodity. Entities that import or refine fuel are required to report their RINs to the EPA and acts as a small environmental subsidy incentive to encourage blending of renewable fuel with petroleum diesel.

In our case, the RIN credit can only be achieved by putting the B100 into a tank truck that previously held petroleum diesel fuel because the 0.01 percent fuel left in the tank truck is considered "blending" in the technical definition when it comes to RIN credits. It's this kind of thing that requires all my willpower not to smash my head repeatedly against the wall.

Chapter 22

The recycling center located in my hometown should be shut down. There is no way to possibly justify its existence if you look at its revenues and expenses. For the sake of efficiency, it should be a couple of containers—one for mixed recyclable containers, one for paper, and one for trash. An attendant could be hired for a living wage to unlock the gate in the morning, make sure the right things go in the right container, lock the gate up at night, and the town would save thousands. My tax bill would go down by a few pennies and those who value efficiency or smaller government would be so pleased.

Efforts to bring sanity to the waste-processing infrastructure in my town of thirty-eight hundred have been attempted a number of times before. Rational arguments by fiscally minded citizens with their spreadsheets, charts, and graphs, together with impassioned pleas at a selectmen's meeting or at the annual town meeting happen on a cyclical basis. The true fiscal weakness of my town's recycling center occurs when the global commodity markets take a nosedive, and the "good news" of the revenue generated by all the recycling suddenly isn't such good news. In spite of a remarkably resilient corps of volunteers who do much of the attending of the recycling center, as well as much of the processing, the numbers don't add up, particularly because there are available options that are absolutely less expensive.

I've had these thoughts myself from time to time, particularly when my tax bill arrives. But this myopic view of anything and everything for the sake of economics or efficiency loses sight of the bigger picture. What gets lost in the spreadsheet view of the world is our lives. Our humanness. Our need for connection and purpose. The very essence of this book is to celebrate the social gathering point that is the town dump—whether it be the one in the town I live, or the one in the city I work in, or the thousands of others all across the country. In evaluating economics how does one quantify the external cost of isolating the community bit by bit? I would argue that saving a couple pennies off the tax bill by streamlining the recycling center to its lowest common denominator would make the community considerably poorer.

We are at our best and our worst at the town dump. The detritus of our lives is exposed to all. Yet there is an earnestness that pervades those who trek to this town common. There you see your friends, neighbors, dentist, lawyer, teacher, political official, out of uniform, doing a chore common to almost all of us, and it binds us in ways that make the affection towards the dump and its experiences palpable. Which is why attempts to curtail the activities at the dump get shot down, vociferously.

Now there's nothing wrong with striving for efficiency and keeping a mindful eye on expenses and revenues. It's not like the dump should be ground zero for anarchy. Quite the opposite, in fact. It's a critical piece of our social fabric, a part of our social contract. We need roads to travel on, we need schools to educate our children, and clean water to drink. We need art and music to raise the human spirit, and we need forums to gather, common purpose to bind us and to rally around the idea that we're all in this together. We have the equalizer that we have by-products of our biological selves. Everyone eliminates, everyone urinates, and the material goods we consume need to have the packaging processed in some way. When something reaches the end of its useful life and is to be discarded, it will be composted or recycled, burned or buried.

My father never specifically told me how he wanted his body handled after he passed away in the spring of 2017 at the age of seventy-nine. He probably didn't tell me because for the prior decade he told me he was going to live to the age of ninety-two, so there was assumed to be plenty of time to have that discussion. I know that beliefs about the proper ritual or disposition of the dead body vary across the religions, but being more spiritual than religious, my overwhelming sense was that my father felt he was the stuff of stardust and while he occupied his human skin he really was a spirit and soul in a human costume. Once his spirit and soul took flight, he left behind an empty vessel and this needed to be managed in some way. My brother and I quickly agreed cremation would allow us to go on tour with him and cast the minerals of his remains in places that were important to him. I also think my father, frugal as he was, would have definitely agreed that the cardboard box that contained his remains was preferable to the costly urns that were offered by the funeral home that cremated his body.

I heard of a company that will package your remains with the proper medium necessary to become part of a planted tree. I'd like my used vessel to become a flaming red in the fall maple tree after my spirit and soul moves on. Maybe I could also be tapped in the spring to provide syrup for some delicious blueberry pancakes, and when I leaf out I could provide shade for all types of life that needs a respite from the hot sun.

I'm quite confident that before I do move on that we will be handling our waste in ways that are difficult conceive in this moment in time. Just like it was hard to imagine the efficiency that the materials-recovery facility that was built in Keene in 1994 would realize versus the workers bending down and picking up a single container at a time in the bulging two-car garage that served as a recycling center prior to 1994, the recycling center of the near future will make today's modern facilities look primitive. We'll look back on our 25–35 percent recycling rates and laugh at the clumsy way we handled our discards over the past thirty–forty years. There have been glimpses of what the future holds. Single-stream recycling plants that feature optical and mechanical sorting systems allow containers and paper to be mixed and transported to the recycling center, creating a transportation efficiency. The most progressive communities have recently added organics to the collection routes and the recycling rates in those communities tick up, although at a fairly steep cost.

A little ways down the road, we will put all our discards in one bin. The one bin will be collected by one truck and taken to a facility that will consolidate material from within a certain radius, perhaps fifty miles depending on the population density within that radius. There, the material will be processed through optical, density, mechanical, and human separation. Fifty-plus sort stations will break down the components of the waste stream into individual components. Those individual components will comprise the raw material for manufacturing new products on the same "campus." Paper recycled through this facility will become new paper right on site, same for a host of plastic products, and on and on.

A quick note on plastics: plastics and their disposition could easily be the subject of another book. For future reference, the issue of micro plastic is just starting to be discussed. In my view, it will be right up there with the biggest environmental issues we face. Now that we know that microscopic particles of plastic are entering the food chain, my guess is this is not a good thing for human health prospects, never mind how plastics in general are impacting ecosystems. This issue of micro plastics will become a bigger deal shortly down the road.

Recycling rates will exceed 90 percent, material that requires burial will be so minimal as to make landfills practically scarce. Much of the material from the waste stream will be recycled, some will be composted, some might go through a pyrolysis process to extract liquid fuel, some will be reused for different applications, and a small portion might be burned for energy. The savings in collection alone will be significant. Recycling rates below 90 percent will be a source of scorn. Much of the energy inherent in the original product will be conserved, and people will no longer be vexed about what to put in the recycling bin. The simple act of tossing an empty can or plastic container, magazine or fully squeezed tube of toothpaste will be an act of conservation. Willful ignorance of discarding our discards can be on an equal playing field as the most ardent environmentalist. The economic hawks will cheer, and rainbows and butterflies will erupt from our chests due to the sheer joy of accomplishment. I can't guarantee the rainbows and butterflies part, but everything else needed for the rest of it is knocking at our door. Some community will take a leadership role and show the rest of us the way. After the first domino falls, communities will be lining up to be next. These new waste processing facilities will be potent economic drivers providing jobs and products to sustain our lives.

I've always appreciated the career I've had, and some exciting challenges lie ahead. Every once in a while, I'll head to the dump after spending my day at the Public Works Department approximately seven miles away. I still oversee the Solid Waste Division in the city of Keene, but for the past twenty years I've served as the assistant public works director for Keene, and in addition to the Solid Waste Division, I oversee the Highway and Fleet Divisions. We all need characters in our lives to make life more interesting, and I've been blessed with an abundance of them.

It is the quiet moments when I'm wandering around the grounds of the dump, the turkey vultures eyeing me like a ribeye steak, the majesty of Mt. Monadnock looming in the distance beyond the grass-covered, sheep-roamed landfill, that I have the chance to reflect on where I've been, where I am, and where I'm going. Inevitably, I pause before getting back in my car parked next to the glass bins. Glass bottles are within easy reach. I pick up a couple, set my feet on the pitcher's rubber and enter my windup. The first pitch is right down the middle—POW! The bottle explodes against the concrete wall shattering into a million pieces. Strike two and three follow in quick succession. The grin on my face says it all. Sixty years old and I'm still getting paid to break bottles!

Epilogue

Perhaps Duncan Pepper Watson was destined to work in the public works field. One of my earliest memories was the curiosity at my initials, DPW, found in every utility hole around town (also known as a manhole, but hey, I'm trying). When I first began my career at the Department of Public Works, DPW, my then mother-in-law gifted me monogrammed golf shirts. I'm not much of a monogram shirt guy, but DPW as a monogram, given my profession, was amusing.

In 2022, I was invited to participate in the New Hampshire Solid Waste Working Group formed by a legislative action to assist the New Hampshire Department of Environmental Services with planning and policy initiatives related to solid waste management. The working group is comprised of members representing various public and private entities involved with solid waste management. I believe the people involved in the working group are well-intentioned, but the chances of any meaningful action coming out of this group are slim. Why? Because too many members of the working group, who have subject expertise, are more interested in maintaining the status quo than in coming up with any solutions to move the needle. Par for the course on this type of working group was working with the Department of Environmental Services to update the state's Solid Waste Plan which is effectively the guidance document for state policy. The working group had several months of meetings, breaking into several sub-committees all tasked with making recommendations of items and issues that should be included in a Solid Waste Plan. Eventually, the state drafted a plan which barely included any of the working group's ideas, and even though the public comments on the draft plan overwhelmingly criticized the plan as being about as far from a plan as a plan could be, the input of the working group and subsequent public comments were largely ignored, and the Solid Waste Plan was submitted to the legislature by the Department of Environmental Services. I made my dissatisfaction with the draft plan known through the working group, then subsequently through the public comment process. This is what I wrote:

> *As the current Assistant Director of Public Works/Solid Waste Manager for the City of Keene, current member of the NH Solid Waste Working Group, former member of the NH Waste Management Council, and former member of the Northeast Resource Recycling Association Board of Trustees (including President of the Board for over a decade), I appreciate the opportunity to provide public comment on the Draft New Hampshire Solid Waste Management Plan R-WMD-22-03.*
>
> *I was eager to be included in the NH Solid Waste Working Group as I saw it as an opportunity to change the course of waste management practices that have largely been static for the past several decades. When I started working for the City of Keene in 1992 recycling was largely aspirational as landfilling handled upwards of 99 percent of the waste stream and our efforts to recycle portions of the waste stream were stymied because we simply lacked the infrastructure to divert any meaningful portion of the waste stream. The City of Keene recognized the importance of infrastructure by investing millions of dollars into constructing a materials recovery facility (MRF), and a transfer station. In 1994 the City of Keene opened what remains the largest publicly operated MRF in the State of New Hampshire, and our diversion rates went from less than 1 percent in 1992 to approximately 25 percent within a year of opening. That diversion rate has stayed relatively the same over the past twenty-eight years.*

In 2022, we saw 75 percent of Keene's waste placed in trucks and transported ninety-two miles away to be disposed of in Waste Management's Turnkey Landfill in Rochester, NH. This status quo will remain for the foreseeable future unless the public and/or the private sector invests in the available existing infrastructure technology that could easily turn Keene's 25 percent diversion rate into a minimum of 75 percent. Without new and expanded infrastructure from public and/or the private sector there will be NO *meaningful progress to the stated goals of this plan (for context the State Plan was to achieve a 25 percent diversion rate by 2030, and 45 percent by 2045). At the bare minimum the plan should be for the Department of Environmental Services, through legislative authority, to facilitate the development of the necessary infrastructure to achieve the metrics of the Solid Waste Plan.*

I would also note that there is a distinct lack of urgency with regard to the stated metrics of the Solid Waste Plan. It was in the early 1990s that the State of New Hampshire had a goal of 40 percent diversion by the year 2000. That goal, and the goals currently included in the 2022 Solid Waste Plan have absolutely no chance of being met because the plan lacks any concrete means of achieving them. Given technology/equipment that is available today the State should be much more aggressive in setting diversion goals of a minimum of 75 percent by 2030. Due to the effects of climate change, and the significant carbon emissions represented by the management of solid waste, it is incumbent on the State to understand the threat of climate change and to take aggressive practical measures to meaningfully reduce carbon emissions associated with waste management. Climate scientists have been sounding the alarm for years now, and people have difficulty understanding the full weight of the coming changes, but within this century temperatures changes will be within a range of 2–6 degrees C. It is impossible to overstate the catastrophic *impacts of even the low end of these temperature ranges, but suffice to say that we do not have the luxury of being passive by setting laughably low diversion goals thinking we have an abundance of time to achieve these inadequate goals. If you're still not convinced about climate change consider this- in 1980 the average interval between $1 billion storm events was 86 days. Today, even accounting for 1980 dollars, the interval is 18 days. Sure, there has been increased development in vulnerable areas since 1980, but that surely does not explain all of it.*

The good news is that we have an opportunity to ask for what we want. If the State of New Hampshire Solid Waste Plan includes a goal of a minimum diversion rate of 75 percent by 2030 it has the potential to unleash innovation through public/private partnerships to achieve the goal. I would stress, however, that the goal cannot be passive. Words such as "shall" or "will" achieve a minimum of 75 percent diversion from landfilling/incineration must be supported legislatively. The public sector, such as the City of Keene, sees 25,000+ tons of solid waste transported to a landfill each year costing millions of dollars in disposal fees, and it's frustrating to know we can do much better. Keene's volume of waste on its own isn't enough to financially support infrastructure to change the status quo, but collectively there is more than enough volume within the State to support the construction and operation of an Advanced Materials Recovery Facility (AMRF) that doesn't skim off a few items typically sorted at a MRF, but rather sorts the entire waste stream into highest and best use diversion.

An AMRF can be sited and constructed by the private sector and the capital cost of construction would be built into the tipping fees that have the likelihood of being equal to or better than current tipping fees while diverting a significant percentage of the waste stream. An AMRF would also serve as an economic driver through creating jobs and value-added manufacturing to resources that currently cease to have value after they are buried.

I understand that many people cling to the idea that education, source separation, waste bans, extended producer responsibility, etcetera, will solve our waste management issues. Within these tools lie some potential; however, when considering the scale of the volume of waste generated in New Hampshire, it is new infrastructure that puts actual resource recovery and diversion at the core that is necessary, and in fact imperative, to affect our shared desired outcomes.

I appreciate the difficulty of the NHDES position to produce a plan given how New Hampshire has historically managed its solid waste. This is complicated by the fact that approximately 50 percent of disposal capacity is consumed by out of state waste. I was pleased to learn at the August 2022 NH Waste Management Council meeting that the NHDES considers the New Hampshire Solid Waste Management Plan goals for diversion to include out of state tonnage. This was new information to me, and provides another opportunity for New Hampshire to do something bold by treating any waste, whether it comes from New Hampshire or out of state, as a resource to manage rather than something to get rid of. I believe we all want the same outcome- to preserve and protect our environment to the greatest extent possible. As constructed, the New Hampshire Solid Waste Management Plan falls considerably short of what is possible and necessary. If this is to be the guidance document for the next ten years, it is no better than the plan that came before it, and not much of anything will change. I urge the NHDES to fulfill their statutory obligation to submit a plan by October 1 with the caveat that before the end of 2023 the plan will be updated with a specific and concrete action plan to achieve a 75 percent diversion goal by 2030. This is New Hampshire's version of a moon-shot. Imagine what people thought in the 1960s about putting people into space and even landing on the moon. Now imagine the year 2030 and the infrastructure to reduce carbon emissions, divert 75 percent of the waste stream, provide manufacturing jobs, and increasing the efficiency in the way waste is collected, transported and managed is in place. New Hampshire would be seen as a leader to the Nation, and this solution exists, right here, right now. It's time to be bold.

I wish I could report that something good came out of my efforts, but we face a collective peril that might just be too late to address. There is a relatively small, but growing chorus of climate doomsayers that posit that the carbon inputs already placed into the environment will affect warming of more than 4 degrees C by 2050. To put this into context, warming of over 3 degrees C is effectively a human extinction event. Even if all human carbon emissions were to magically cease today, the catastrophic warming will still occur. Pretty sobering, particularly when too few seem to acknowledge or even care. Yes, climate change is a natural cycle. One that usually unfolds over centuries. That humans are accelerating this natural cycle is supported by overwhelming scientific evidence. The only remaining question is the timeline for this cataclysm to unfold. As I type these words, it is dumping heavy snow outside my window. Hardly the stuff of "global warming," but lumping weather, which is what we see on a daily basis, versus climate, which are the longer-term patterns, is the cornerstone of climate denial. Why so many are willing to believe in a smoky deity in the heavens guiding our lives, but don't believe in climate change will always be a mystery to me. But the direction the science indicates we're heading challenges any embers of optimism I have. I waffle between the feelings of the Prince lyrics from his hit *1999* (look up the lyrics, because I can't reprint them here due to copyright laws, but the gist is the party is over, and we're out of time) and the stubborn determination that a possible technological fix might exist. The mere fact that I mention Solar Radiation Management as a potentially "viable" technology reflects my desperation for something, anything really, to give me a sense of hope. Climate engineering whereby solar radiation (sunlight) is reflected back into space, mitigating much of the warming associated with climate change seems reasonably simplistic until you consider the scale that would be required for a global impact, not to even begin mentioning the infinite unintended consequences. But, we're in throwing spaghetti against the wall time to find the idea or ideas that might work. It's difficult to fathom the critical mass of human cooperation necessary to make global endeavors work, considering the challenges we faced getting humans to unite around public health policy during the Covid pandemic.

Climate migration will be the tipping point. We see the beginnings now, although much of human migration now is due to economics versus inhabitability. Whether it be rising sea levels, drought, severe weather, or a combination of these things, humans are going to be on the move in the decades ahead like we've never seen before. I hear news reports of upwards of two million immigrants attempt to cross our southern border each year, but what's going to happen if we have twenty million or two hundred million people on the move? Even within our own country, we are likely to see large cities unable to supply drinking water to their residents. With rising seawater, or no access to potable water, or severe weather destroying homes, people will want to move to areas that are not threatened by rising water, have access to water drinking, sanitary services, and agriculture, and are less prone to severe weather. Maybe you live in one of the affected areas now. Maybe you live in a relatively climate resilient part of the world. Those people who are forced to relocate will want to relocate to those areas where the prospects are better. Will the people that live in the habitable places welcome their brothers and sisters with open arm? Our human history would suggest that we will not.

About the same time I was feeling at my most cynical about the human experiment, the pandemic continuing to wreak havoc, healthcare, particularly mental-health systems stretched too thin, political fallout from a nearly successful overthrow of democracy, and rumblings of war in Ukraine, I was contacted, out of the blue by Melanie Kohn. Melanie and I had first met in 1974 during the production of *Be My Valentine, Charlie Brown*. Melanie voiced Lucy. Ironically, her sister had the role prior to her selection, so Lucy was definitely in her family. My hazy memory of her was a figment of an imagined "big" girl (and by "big," I mean older and taller than me) who stole my candy bars. My personal urban legend of Melanie was that she was the actual embodiment of Lucy, the irredeemable bully who tormented Charlie Brown by ALWAYS pulling the football out from under Charlie Brown as he attempted to kick it, sending him flying through the air. As a quick aside, apparently my yell recorded from the football trick was noteworthy enough that I got a call from an ad agency fifteen years ago asking if they could use the "Auuuuuuuuugh" yell for a commercial they were making. I don't actually recall what the product was that they wanted to use my voice for, but I do recall them saying they were willing to pay me six hundred dollars for the two-second sound clip. I just started laughing. The woman on the other end of the phone asked why I was laughing and I said, "How would you respond if you got a random call from someone asking to use a two-second clip of your voice in something you recorded thirty years prior and wanted to pay you six hundred for it?"

The truth of Melanie could not have been farther from my imagination. Initially, I didn't share my hard-wired impression of her because she seemed very pleasant, and she was calling with a proposition. Hmmmmm, a proposition from Lucy? Was I going to fall for the football trick again? Not this time! Instead, she was calling about comic conventions. She had recently been attending comic conventions as an invited guest, and she wondered if I might be interested in joining her. I asked her for a couple weeks to think about it, then got back to her, without fully understanding what I was signing up for and agreed to join her for select shows.

I had no idea what a truly marvelous world Melanie was giving me access to. In the Spring of 2022, I attended my first comic convention in Iselin, NJ. With Melanie's guidance, I acquired some Charlie Brown collectables, found some good images for the #1 item necessary for a voice actor—the 8" x 10" photo. I was curious why anyone would want an 8" x 10" photo of me, but Lucy stepped in and said, "Not of you, you blockhead. The 8x10s need to be Charlie Brown." *Ahhhhhhh*, now I get it. Usually, as a comic convention guest, you are provided with transportation and lodging costs, and a guaranteed minimum amount that you will make at your table. If for some reason the sales of the collectibles and 8x10s don't add up to the guaranteed minimum, the convention organizer makes up the difference. I was determined not to fall into the trap of the Charlie Brown of my expectations being in a different universe than reality.

The morning of the show, I set my eyes on Melanie for the first time in forty-five years. It was a lovely reunion, and I relied heavily on Melanie throughout the day, as I didn't know how anything worked, while she by then had several shows under her belt and was making the comic convention scene her livelihood. I wasn't about to quit my day job, at least not yet, but what I found endearing about Melanie is that she was experiencing the same wonder at the circumstances that would lead us to be in the same room at the same time representing something we did more than two score years before.

At that particular show, which was billed as a "*Peanuts* Reunion," we were also joined by Patricia Patts, who voiced Peppermint Patty, but not in any of the shows that Melanie and I had done together. Patricia had also been one of the original Annies in the Broadway tour of "Annie." Honestly, I just couldn't believe that anyone would be interested in me or something I had done so long ago, but when the doors opened I couldn't have been more wrong. To give you an idea of how iconic Charlie Brown—and by extension the voice attributed to the character—is, I recently received this letter in the mail:

Dear Mr. Watson,

I want to start off by thanking you for reading my letter and that it has reached you in good health.

I have been a fan of yours for as long as I can remember. I remember sitting on the couch with my brother and older sister, watching your movies and shows with our Mom. She adored you up until the day she passed and I am sure she adores you still! One of my Mom's favorites is Race for Your Life, Charlie Brown! *Every time I watch it, it brings back memories of her. Some of those memories bring back smiles and laughter, others bring back tears, but I enjoy the happy memories much more!*

Our Mom (Claudia), had dreams of becoming an actress when she was younger, but when she found out she was having my older sister, taking care of her took precedence over everything else. As we got older, she would act for us and talk about how much she still admires you! I think . . . let me correct myself. . . . I KNOW she would have made a wonderful actress, but she was an AMAZING mother! Mom was diagnosed with Parkinson's Disease several years ago, it progressed so quickly, so she no longer had control of her motor skills and eventually no longer knew who we were anymore, but we never forgot who she was. There were times when she regained memories, and some of those memories included you! She talked about you as if she has known you her entire life, and in a way, she has from watching your films!

We have become such big fans of yours and we will be beyond grateful if you will sign the photos I enclosed with my letter for us. I have picture frames ready for them and I am going to surprise my brother, Frank, and my sister Amanda, along with one for myself.

It has been a delight writing you. I hope you and your family are doing well and living life to the very fullest!

All My Best,

Tom

Melanie is always quick to point out that my character was and is the franchise. There have likely been many arguments on this topic, and probably, if asked, Snoopy would be the most beloved *Peanuts* character given a proper sample size. But what makes *Peanuts* work is Charlie Brown. The ultimate optimist who gets knocked down, time and time again, only to get up and persevere. I am an introvert. And an introvert representing one of the all-time icons in cartoon history is a perplexing position to find oneself in. Yet when the doors open and people walk by your booth, stop for a moment to ponder what they are seeing, then excitedly come up to share a memory or to purchase a keepsake, I find myself out of my body, inhabiting Charlie Brown as I did when I was twelve.

One of the epiphanies I've had, and this has been reinforced through attending a half-dozen comic conventions the past year, is that these gatherings are ground zero for nerds and geeks. I use the words nerds and geeks in the least derogatory fashion possible. Nerds and geeks are awesome. Now I didn't think nerds and geeks were awesome when I was twelve. I wanted to be a cool kid. I never made it to the cool kid crowd (thankfully) and had fended off every invitation to come to the nerd and geek side. As a consequence, I found myself in neither world, but largely alone (the exception being my best mate, Michael, that I recently reunited with on a trip to California courtesy of an invitation from the San Francisco Giants to be part of *Peanuts* Night, and my role was to say "Good grief . . . play ball" into the microphone for the thirty-five thousand people in attendance). I wish I knew then what I know now—that nerds and geeks are the most interesting, positive, fabulous freak-flag flying, inclusive, and accepting people. And while nerds and geeks make up the vast majority of my fan base, I, in turn am a fan of nerds and geeks of every color, size, and shape. I'm so grateful that they have been patient with me while this realization fully blossomed.

Most of the shared memories are along the lines of the letter I received from Tom, but people also want to know other details- a partial list of Q's & A's follows:

Did you ever meet Charles Schulz?
I never had the pleasure. I did write to him when I was voicing the character asking to meet him, but I received a nice letter from his secretary indicating that he was too busy to schedule something. I do know that several of the voice actors interacted with Charles Schulz, but Melanie and I did not ever meet him. We did, however, meet and interact with Lee Mendelson and Bill Melendez. They were the executive producers of the TV specials, and lovely men. Bill also was the voice of Snoopy and Woodstock, making these wonderful sounds come out of his mouth and bringing them to life. I did, however, recently meet Craig Schulz, son of Charles Schulz, and his lovely wife, Judy, as part of the San Francisco Giants *Peanuts* Night.

Did you work with the other child actors?
Yes, on occasion. We were all in the studio for a group sing-along *She'll Be Coming Around the Mountain*, which was featured in *Race for Your Life, Charlie Brown*. More often though, we were in the studio alone with headphones and a script and we were directed via Lee or Bill in the control room, asking us to read a line a certain way or to put in more or less inflection.

What are your favorite memories of being Charlie Brown?
One that stands out, aside from the HUGE (for a twelve-year-old) checks, is when the script called for me to sigh. I kept reading the word instead of making the sighing sound which is what they wanted and it took several minutes for me to understand they wanted a sound, not a word.

The other memory is going to provoke a memory for many, and a collective shrug for anyone of the internet age. There was once a publication far more popular than the Bible, that people consulted every day. It was known as the *TV Guide*. If you wanted to know what was on television you consulted the *TV Guide*. There were no VCRs, no DVDs. If you wanted to watch a particular show you had to be on the couch at the prescribed time to watch. I was over the moon the day my mom brought home a *TV Guide*, and there, in black and white, was the listing for *Be My Valentine, Charlie Brown* and right underneath the listing was my name credited as the voice of Charlie Brown. I cannot overstate how cool I thought that was.

Did you ever do any other acting work?
Close, but no cigar. I was cast as a lead in what eventually became a box-office smash family film involving a boy and his dog, but when the shoot was to begin, I was unreachable due to being on a lengthy backpacking trip with the Boy Scouts, so they picked someone else to play the lead. But, if anyone has a vision to create an animated version of adult *Peanuts*, or something else involving voice acting, give me a shout!!

Do you still get residuals?
Unbelievably, yes. When I was first selected to be Charlie Brown, I had to join the Screen Actors Guild. Best investment I've ever made. They track all of that stuff, and every once in a while a check will arrive in the mail. The checks are not frequent, and more often than not they are for less than twenty dollars, but imagine what it would be like to be walking along the street and seeing a fresh twenty-dollar bill just lying there waiting for you to pick it up. That's what getting a residual check is like.

What do you think happened to the Peanuts characters when they grew up?
Pretty sure Charlie Brown would have grown up and had a multi-decade public works career. Charlie Brown is all about public service.

Lucy became a therapist, then later a TV therapist where her cure for everything is "Just get over it."

Schroeder was a classical pianist, but later a keyboardist for a heavy metal band.

Linus started a Great Pumpkin theme park to keep the magic of believing alive for generations to come.

Sally still does slam poetry shows along with creative writing for the *Atlantic* and *New Yorker*.

Peppermint Patty and Marcie, come on! Do I really have to spell it out for you? Together they run an organic farm and consistently win prizes for their farm products at regional fairs.

Anything else?
I learned, via my getting to meet Craig Schulz recently, that the *Peanuts* franchise is alive and well, with planned features on Marcie and Franklin. I also learned that *You're a Good Sport, Charlie Brown* was based on Craig Schulz's motocross racing days, and that particular TV special won an Emmy. Pretty cool to think I had a part in an Emmy-award-winning show.

I'm often asked about my favorite Charlie Brown line and, really, how could it be anything other than "Good grief"? It's as versatile as a cuss word, but can be uttered in almost any circumstance, clearly conveys an emotion or feeling, and I find it has worked its way into my lexicon more and more each passing year.

And one very little-known fact—I also voiced Franklin in *You're a Good Sport, Charlie Brown*. According to Peanuts.com, Franklin is a busy kid: he plays baseball and is learning guitar, he's a member of a swim club and of 4H. Though his life is active, Franklin is never too busy to help his friends. He is supportive and smart and always willing to lend a hand. He also enjoys spending time with his grandparents and learning about *the old days* from them. And even though he thinks Charlie Brown's friends are a little weird, he's happy to be part of the gang.

I will admit that it's a little strange to have voiced the first African American character in the *Peanuts* gallery, and I only found out about it by happenstance. I really wasn't even aware that I had voiced his character until earlier this year, when someone pointed out that I'm the credited voice actor. Apparently, it was somewhat common for the child actors to voice other characters in the shows because they usually had just a few lines, and it was more efficient to use the actors on hand than to make the supporting characters have their own voicing. Whatever the reason, I hope I voiced Franklin well.

About the Author

I definitely fall into the "work to live" category of persona, and while I have dedicated a significant portion of my life to the service of the citizens of Keene, I also had a full life outside of work. I ski, I hike, I bike, I teach yoga. I garden, I play guitar, I cook up a storm. I'm the father of lovely girls, I've been the human to memorable canines. I'm the son of remarkable parents, I'm the brother of an outstanding man. The bulk of my work life has been spent working for the City of Keene Department of Public Works, and before that I stumbled into a gig as the voice of Charlie Brown in several Peanuts TV Specials (*Be My Valentine, Charlie Brown*, *You're a Good Sport, Charlie Brown*, and *Happy Anniversary, Charlie Brown*), and one movie (*Race for Your Life, Charlie Brown*). I've loved deeply and been loved deeply. I have friends in high and low places, and humor is often what gets me through a day.

Because I am a human, I leave a footprint of my existence. Collectively, as a species, we need to lighten that footprint if humanity is to survive. To me, the best experiences in life result in stories. I'm eternally fascinated by the sociology and psychology of human behavior when it comes to the disposition of resources when they've reached the end of their life. You say discards, trash, refuse, and waste; I say resources. We can absolutely do a better job treating discards as resources, and increasingly I think we will. Our collective fate will, in part, depend on it. I'm certain you have your own stories that involve trash. Everyone does. These are my stories, and I hope you enjoy them.